IB Geography

Option D

Hazards and Disasters:

Risk Assessment and Response

Revision Guide

Garrett Nagle

Ben Tavener

About this book

The purpose of this book

This revision guide was written to provide a concise revision resource covering each individual syllabus statement for IB Geography optional theme D: Hazards and Disasters: Risk Assessment and Response. Throughout this book, we have included a range of different contemporary "case studies" which can be read and learnt to help maximise the effectiveness of your IB exam answers.

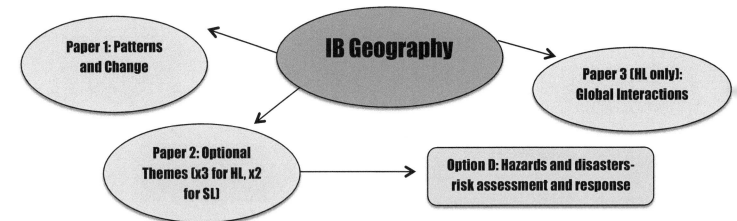

How to use this book

We recommend that you use this guide as an additional revision "tool" and encourage you to use it in conjunction with other resources, such as your textbook, class notes and hand-outs. You should only use this book after you have covered this optional theme in class. Do not use this book for "self-teaching" as no revision guide will ever be able to replace the value of a fully qualified Geography teacher.

After you have read and understood the content of this guide, you should try to practise some of the IB exam focus questions, which have been written in the IB exam style. In addition, we have provided advice on how to tackle these, question by question. Throughout this revision guide you will find several exam hints, which have been written by an experienced IB Geography examiner. These hints outline common errors that students have made in previous examinations and give you tips on how to use the material given within this guide to answer actual IB exam questions.

Good luck with the revision for your exams and we hope this guide proves to be useful in the process!

Garrett Nagle & Ben Tavener

Acknowledgements

For Angela, Rosie, Patrick and Bethany - for their continued support, patience and good humour.

Garrett Nagle

Warm thanks must go to Dr Garrett Nagle for offering support and advice whenever it was needed. Thanks must also go to Mr James Cope, Aimee Mowat-Helling, Flora Cameron Watt, Lucy Poffley and Oliver Ford for their support and enthusiasm. Additional thanks are given to Charlotte and Rachel Tavener, Alec Lingorski and Stuart Wilson. I am very grateful to Shutterstock® for granting us kind permission to use the photos that appear in this guide and to Martin Sanders at Beehive Illustration®. Final thanks must be given to Karen Tavener, who helped in more ways than I can list here.

Ben Tavener

Contents

Section 1: Characteristics of hazards

What are the characteristics of hazards?

Key definition!

❖ A hazard is a threat (whether natural or human-induced) that has the potential to cause socio-economic (e.g. loss of life and property) and environmental (e.g. loss of habitat) damage.

❖ A hazard is a potential threat and a hazard event only occurs when a geophysical event occurs in an area with a vulnerable population (e.g. slum settlement). For example, if an earthquake were to strike in the middle of an uninhabited and biologically unproductive desert region, the occurring earthquake would be classified as a hazard as it has the potential to cause damage, but has not caused socio-economic or environmental damage.

❖ You have studied **two** main types of hazards: **natural hazards** (e.g. earthquakes, volcanic hazards, hurricanes and droughts) and **human-induced hazards** (e.g. chemical explosions and oil spills).

❖ The spatial distribution of hazards is widespread. This is because hazards can occur on any scale and form part of our everyday life. However, from the map below you can see that the distribution of individual hazards (earthquakes, volcanic hazards, hurricanes, droughts and human-induced hazards) is much more limited as their occurrence requires specific physical or human conditions.

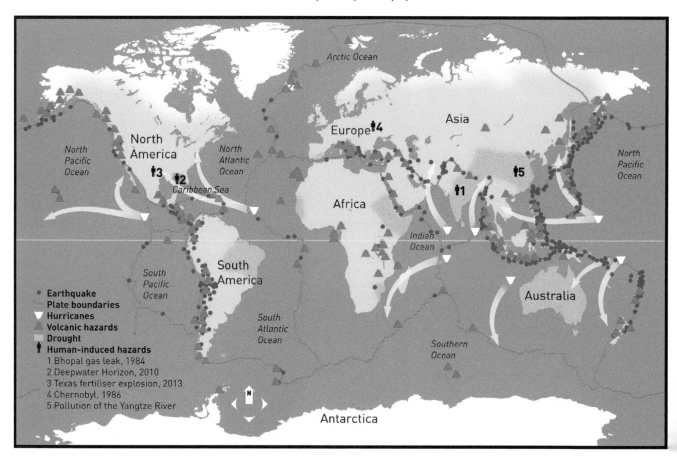

Figure 1: Map showing the spatial distribution of hazards.

We can compare the characteristics of hazards by examining the following variables for each hazard. However, it should be noted that the characteristics of each individual hazard event vary and do not always follow the typical characteristic profile.

1. **Spatial extent:** The size of the area affected by the hazard.
2. **Predictability:** The measure of how predictable a hazard is.
3. **Frequency:** How often a hazard occurs/its distribution through time (i.e. its return period).
4. **Magnitude:** The size of a hazard event on a scale (e.g. Saffir-Simpson Hurricane Scale).
5. **Duration:** The length of time that the hazard event lasts.
6. **Speed of onset:** The time difference between the occurrence of a hazard and its impacts.
7. **Effects:** The short-term/long-term socio-economic and environmental impacts of a hazard event.

IB Geography **Hazards and Disasters: Risk Assessment and Response**

What are the characteristics of earthquakes?

❖ An earthquake is a sudden movement of the earth's crust that can be caused by the release of accumulated stress mainly along fault lines, plate boundaries and volcanoes.

The spatial distribution and origin of earthquakes

❖ The distribution of earthquakes is far from random as they occur in distinct regions across the globe. Most earthquakes occur around the Pacific Ocean or "Ring of Fire", the mid-Atlantic ridge, in south-east Asia and parts of south-east Europe.

❖ Most earthquakes occur in these regions because they are situated along the edge of the world's tectonic **plate boundaries**. When plates converge and collide, a large amount of tension is released in the form of shock waves, causing an earthquake to occur. Less destructive earthquakes occur at **constructive** plate boundaries because less friction is caused by plate divergence. However, **transform faults** still exist here (e.g. Mid-Atlantic Ridge).

❖ Seismic activity can also occur within close proximity to volcanoes. This is because when volcanoes erupt, large amounts of molten rock are forced upwards, causing large amounts of pressure to be released and large tremors to occur (e.g. Chile, February 2010).

❖ Some earthquakes occur a large distance away from plate boundaries (e.g. UK). These are usually explained by fault lines (cracks) within the surface rock or human activities, such as dams and fracking, which put large amounts of pressure onto the underlying rock.

The characteristics of earthquakes

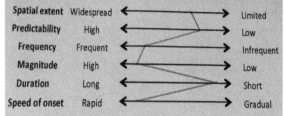

1. **Spatial extent:** Areas close to the epicentre experience the worst of the impacts. The intensity of the earthquake reduces as the distance from the epicentre increases. Earthquakes of a larger magnitude have a larger spatial extent, as stronger seismic waves are able to travel further.

2. **Predictability:** Scientists are able to use probability (historic trends) and methods such as groundwater movements and knowledge of geology, to predict the rough location/frequency of an earthquake. However, it is difficult to predict exactly where an earthquake will occur along a fault line, when it will occur and its magnitude.

Figure 2: A typical earthquake hazard profile.

3. **Frequency:** It is estimated that several million earthquakes occur each year. However, only about 20,000 of these are of a detectable magnitude.

4. **Magnitude:** Most earthquakes are of a moderate intensity. Earthquakes of higher magnitude are less frequent than earthquakes of a lower magnitude. The magnitude of an earthquake is dependent on the amount of physical displacement along the fault-line.

5. **Duration:** Most earthquakes have a very short duration of just a few seconds. However, the aftershocks can shake affected regions for several weeks after the initial earthquake.

6. **Speed of onset:** The onset of an earthquake is relatively rapid because seismic waves move very quickly up to and across the surface of the ground. However, it is important to note that it is the slower secondary (S) waves that cause the most impact as they vibrate at right angles and hence make the ground move horizontally.

> **Exam hint!** It is important that you display an awareness of different physical and human factors influencing the intensity of hazard events. For example, for earthquakes physical factors include the magnitude, geology, depth of focus, peak ground acceleration and duration, whereas human factors tend to revolve around preparedness levels and quality of housing, buildings and emergency services.

Effects of earthquakes

Primary effects

- **Ground shaking:** The movements of the ground caused by the seismic waves can fell buildings, bridges, trees, etc. killing and trapping people.

Secondary effects

- **Soil liquefaction**: Soils with high water content lose their mechanical strength when violently shaken and start to behave as fluid. This therefore allows buildings and other objects to sink into the soil.
- **Landslides/avalanches**: The sudden shaking of the earth can weaken the structure of rocks and surface soils, causing landslides and avalanches to occur (e.g. El Salvador, 2001).
- **Tsunamis**: When earthquakes occur underneath the seabed, a large volume of water can be displaced forming a giant sea wave. Tsunamis result in large-scale coastal flooding and can cause more damage than the earthquakes that trigger them (e.g. Indian Ocean, 2004 and Japan, 2011).
- **Fires**: The initial shaking of an earthquake can rapture gas cables, triggering uncontrolled fires (e.g. Kobe, 1995).

Case study: Haiti earthquake, 2010

On 12th January 2010, a magnitude 7.0 earthquake struck 25km west of Port au Prince, Haiti. The earthquake originated at a destructive plate boundary, where the Caribbean plate sank beneath the North American plate. The focus of the earthquake was only 13km beneath the surface and so the impact of the quake was severe. The earthquake was also felt in surrounding islands, such as Cuba and Jamaica.

As this earthquake occurred in the late afternoon, the effects of the earthquake were intensified, since more people were present in the city centre of Port au Prince. It is estimated that over 316,000 people were killed by building collapse and fires caused by the earthquake. Over a million Haitians were left homeless and a further 300,000 were injured in the earthquake.

Figure 3: Reconstruction for the Haiti earthquake is estimated to take over 30 years.

Secondary effects included damage to communication systems and transport facilities (A main road linking Port au Prince and Jacmel was blocked for 10 days) causing vital rescue and aid efforts to be delayed. In the months following the earthquake, over 40,000 cases of cholera were reported as a result of poor sanitation and the damage and destruction of waste pipes. Furthermore, in March 2010, Haiti received a large amount of rainfall, which resulted in mudslides and the spread of other water-borne diseases.

What are the characteristics of volcanic hazards?

Key definition!   ❖   A volcano is a landform created at weaknesses within the earth's crust, which erupts molten magma (lava) and gases following the release of pressure within the intrusive magma chamber.

The spatial distribution and origin of volcanic hazards

- ❖   The distribution of volcanoes is fairly limited as they can only be located at areas with access to a supply of **magma**. Most volcanoes can be found at destructive and constructive plate boundaries. However, volcanoes can also form at **hotspots**, such as Hawaii, where a weakness in the crust allows magma to rise up from the mantle.
- ❖   At **destructive plate boundaries**, the oceanic plate sinks underneath the continental plate causing the oceanic plate to enter the **subduction zone**. Following exposure to immense heat and pressure, the oceanic plate is melted and destroyed, causing magma to rise up from the subduction zone. Volcanoes found at destructive margins are known as **Plinian-type** volcanoes. These cone-shaped volcanoes are very explosive and produce thick viscous lavas.

❖ At **constructive boundaries**, a gap within the Earth's crust is created when two oceanic plates move away from each other. Magma is then able to rise up through this gap to form a volcanic island in the ocean. Volcanoes found at constructive boundaries and hotspot volcanoes are collectively known as **Hawaiian-type** volcanoes and are characterized as being short and wide and able to produce fluid lavas and hence erupt less violently. The duration of these eruptions are much shorter than Plinian eruptions.

Figure 4: Mount Fuji, Japan- a Plinian (cone) volcano.

Figure 5: Mauna Kea, Hawaii- a Hawaiian (shield) volcano.

The characteristics of volcanic hazards

Spatial extent	Widespread	⟷	Limited
Predictability	High	⟷	Low
Frequency	Frequent	⟷	Infrequent
Magnitude	High	⟷	Low
Duration	Long	⟷	Short
Speed of onset	Rapid	⟷	Gradual

Figure 6: A typical hazard profile for volcanic hazards.

1. **Spatial extent:** The finer products ejected from a volcano, such as ash, can travel hundreds of kilometres away from a volcano. Volcanic ash clouds can also rise up 25 kilometres into the Earth's lower atmosphere (e.g. Eyjafjalljokull, 2010).
2. **Predictability:** Volcanoes give off many useful warning signs that enable scientists to predict the time-scale of the upcoming eruption. Common warning signs include ground tremors, changes in the volcano's physical appearance and the emission of sulphur from the crater. However, scientists find it more difficult to predict the intensity of a volcanic eruption.
3. **Frequency:** Although there are over 500 active volcanoes in the world today, on average only 50 volcanic eruptions occur each year. The frequency of Hawaiian-type volcanoes erupting is higher than the eruption of Plinian-type volcanoes, but the magnitude of these eruptions is smaller.
4. **Magnitude:** The magnitude of a volcanic eruption is dependent on the type and amount of magma stored within the magma chamber. For example, Hawaiian-type volcanoes erupt fluid lavas and hence erupt at a lower magnitude, whereas Plinian-type volcanoes erupt viscous lavas, which erupt at a much larger intensity due to the heavy weight of the lava.
5. **Duration:** The initial eruption of a volcano can vary from a few minutes long to a few years long (e.g. the Soufriere Hills volcano in Montserrat erupted for 5 years from 1995 to 2000) as the duration of an eruption depends on the amount of magma contained in the underground chamber. The effects of an eruption can last for a prolonged period of time.
6. **Speed of onset:** The "time lag" between the eruption(s) of a volcano and the peak intensity of the impacts can be relatively rapid. For example, the pyroclastic flow can kill populations within minutes of the eruption (e.g. Mt. St Helens' Pyroclastic flow was responsible for the death of 57 people within a day of the initial eruption). However, the speed of onset for many of the other serious impacts, such as falling ash and volcanic mudflows, is much slower as they can take hours to cause injuries and homelessness.

Effects of volcanic hazards

Primary effects

× **Ejection of rock into the atmosphere:** When a volcano erupts, fragmented rocks and minerals are ejected into the air. The heavier particles (e.g. volcanic bombs) fall near to the base of the volcano whereas the lighter particles (e.g. ash) are carried by the wind. Ash can fall in settlements far from the volcano causing building collapse and breathing difficulties. For example, ash ejected by the Eyjafjalljokull volcano in 2010 spread across parts of western Europe, causing large-scale air travel disruption.

× **Emission of volcanic gases:** The three main gases released are carbon monoxide, carbon dioxide and sulphur dioxide. Carbon monoxide is dangerous as it bonds with the haemoglobin within the human blood stream and reduces the blood's capacity to carry oxygen around the body leading to fainting and eventually death. Increased carbon dioxide levels can be dangerous as it has a density much greater than air, causing suffocation in humans. Sulphur dioxide can have many detrimental effects on the environment, such as the acidification of lakes and soils and the formation of acid rain.

× **Pyroclastic flows:** These are fast-moving (up to 800km/hr) flows of gases and pyroclastic material, such as ash and cinders. When this super-heated (temperature >500°C) dense cloud moves downhill, it destroys anything and anyone in its path by a combination of internal and external burning.

Secondary effects

× **Mudflows:** Volcanic mudflows occur in steep and tropical volcanic areas, such as Indonesia. These are formed when volcanic material from an eruption is mixed with water from heavy rainfall. These hot and heavy mudflows have the ability to destroy homes and cause agonizing external burns.

Figure 7: Damage caused by a volcanic mudflow, Jogjakarta, Indonesia.

× **Tsunamis:** Both the tremors caused by a volcanic eruption and the amount of volcanic debris deposited in the sea can lead to a large displacement of water (e.g. following Krakatoa, 1883).

× **Volcanic landslides:** When the crater of the volcano explodes, the adjoining land collapses and moves down the slope, causing loss of farmland and housing.

× **Climate change:** The ejection of large quantities of thick ash and greenhouse gases (e.g. carbon dioxide) from a volcano can cause short-term changes in the global climate. For example, following the eruption of Pinatubo in 1991, global temperatures decreased by 0.5°C. Scientists estimate that the overdue eruption of the Yellowstone super volcano could have the potential to cause a 5 year-long volcanic winter, resulting in mass plant extinction.

Case study: Mount Merapi, Java, 2010

Mount Merapi in central Java, Indonesia erupted violently several times throughout October and November 2010. This Plinian-type volcano was formed at a destructive plate boundary, where the Indo-Australian plate subducted beneath the Eurasian plate. There were many signs of the eruption; increased seismic activity, swelling of the volcano and lava flows. On 25th October 2010, the volcano erupted three times causing lava to flow down its southern slopes. The volcano erupted continuously until 30th November 2010.

The immediate impacts of the eruption included the ejection of super-heated volcanic bombs and gases, which spread out over 11 km from the volcano. Furthermore, a pyroclastic flow travelled 3 km down the slope of the volcano, destroying both vegetation and houses. The ash emitted from the volcano fell over 30 km away, causing displaced survivors of the eruption to be affected by medical conditions such as respiratory difficulties and headaches. It has been estimated that 353 people were killed by this eruption. Over 320,000 people were evacuated from the villages surrounding the volcano. The main secondary hazard of the eruption was the mudflows, which injured hundreds and destroyed many homes. The ash and the pyroclastic flow also affected the surrounding vegetation causing the price of essential vegetables, such as potatoes, to increase. Subsequently, many farmers temporarily lost their livelihoods. However, it is likely that this ash fall could improve the fertility of the soil in the long-term. The volcanic ash plumes caused widespread disruption to air travel as the safety risk posed by ash getting into an aircraft's engine was very high. As a result, hundreds of flights into and out of Java's busiest airports were suspended until the ash concentration in the stratosphere had dropped and dozens were cancelled. Due to travel disruptions, the island's income from tourism has been seriously affected. Flights departing from western Australia were also affected by the eruption.

What are the characteristics of hurricanes?

:y definition!   ❖   A hurricane is a severe tropical cyclone that has wind speeds greater than 74 mph.

The spatial distribution and origin of hurricanes

❖ Hurricanes usually form between 5° and 20° north or south of the equator. Most of the Earth's hurricanes are formed in the Gulf of Mexico, the Arabian Sea, the East Pacific and the Bay of Bengal.

❖ Conditions needed for a hurricane to form include an oceanic area with a temperature of above **26.5°C** (needed for evapotranspiration and latent heat), a water depth of at least **70m** (provides sufficient moisture) and a location that is at least **5° north or south of the equator** (stronger Coriolis effect).

❖ In order to understand the complex formation of a hurricane, it is important to be aware of the mechanisms of **ITCZ** and the **Coriolis effect.**

❖ The **inter-tropical convergence zone (ITCZ)** is a band of cloud that covers the solar equator. The overhead sun, and hence the ITCZ, varies according to the tilt angle of the Earth's axis. When the northern hemisphere receives the most sunlight (June-August), the ITCZ moves north of the equator and when the Earth's southern hemisphere receives the most sunlight (December), the ITCZ moves south of the equator, creating a low pressure zone.

❖ Hurricanes rarely form when the ITCZ is over the equator because the equator has a very weak Coriolis force.

❖ The **Coriolis "spinning" effect** is a force created by the Earth's rotation that deflects straight movements of air into curving paths, giving a turning effect.

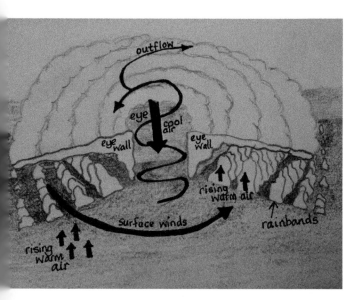

Figure 8: Diagram showing the processes involved in hurricane formation.

1. Hurricanes occur where the **ITCZ** is at its most northerly/southerly extent and when the trade winds converge.
2. During the summer months, the sun heats the ocean water causing the water temperature to rise to above **26.5°C**. As a result of this, the rate of evaporation increases.
3. As this air rises, water vapour cools and condenses into small droplets of water. During this process of condensation, **latent heat** is released, fuelling the storm even further.
4. Eventually, these water droplets collide with one another, become bigger and fall as **convectional rainfall**.
5. Due to the Coriolis force, the whole storm begins to spin around the calm eye. As a result of this, rising warm air is balanced by descending cool air in the centre. The eye of the hurricane has a high pressure and is surrounded by the **eye wall**, which contains the strongest winds in the storm.
6. Over time, the spiralling storm moves across the oceans in the westward direction towards land. However, as hurricanes move over land or cooler oceans they run out of latent heat and **dissipate**.

The characteristics of hurricanes

1. **Spatial extent:** Hurricanes travelling at a high speed can affect a number of different countries. However, the majority of the damage in a particular country will be confined to the coastal areas due to the fact that hurricanes tend to dissipate as they move further inland.
2. **Predictability:** Scientists are able to deduce which areas are most at risk from hurricanes, as they require specific conditions found only between 5° and 20° north/south of the equator. Furthermore,

techniques such as satellite imaging and buoys can be used to monitor the wind speed and direction of a hurricane. However, it is still very difficult to predict the pathway of a hurricane.

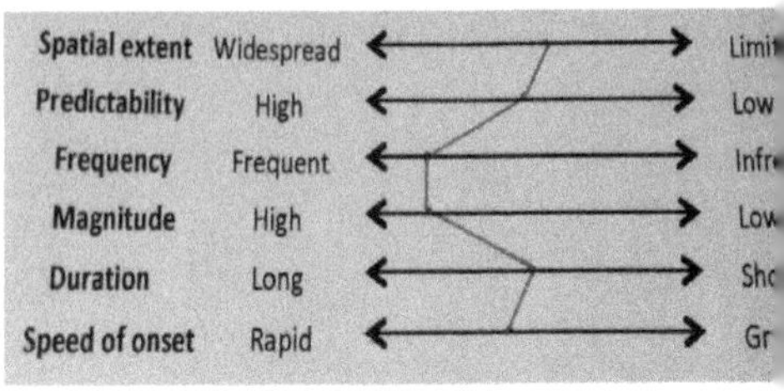

Figure 9: A typical hurricane hazard profile

3. **Frequency:** In the northern hemisphere, over 90 hurricanes form each year with the majority of these forming from August to October. However, some scientists claim that the frequency of hurricanes is increasing due to the effect of human-induced global climate change, which is increasing the number of warm oceanic areas suitable for hurricane formation.

4. **Magnitude:** The magnitude of a hurricane is usually very high because once formed they are essentially self-fuelling as further latent heat is created and re-used from the condensation of water vapour. However, the magnitude of a particular hurricane is dependent on the depth, temperature of the oceanic area and strength of the Coriolis force.

5. **Duration:** The destructive forces of a hurricane can affect one location for a couple of hours. However, the average lifespan of a hurricane is between a week and two weeks. Hurricanes usually last no longer than this because the storm will eventually lose its original source of latent heat when it passes over land or cooler oceanic areas.

6. **Speed of onset:** As hurricanes usually begin life as gradually developing tropical storms, they usually take a couple of days to evolve into an intense hurricane. This large period of time allows communities sufficient time to prepare in advance by evacuation and other strategies.

Effects of hurricanes

Primary effects

× **Strong winds** of up to 320km/hour which can cause the instability of soils (and consequently crops) and severe damage to buildings and local infrastructure (e.g. bridges and power supply).

× **Intense rainfall**: Since hurricanes absorb so much moisture from the oceans, they can result in a landfall of 500 mm in 24 hours.

Secondary effects

× **Storm surges**: Stormy seas that result when the friction between the wind and the surface of the sea creates large waves. These storm surges are dangerous as they result in coastal flooding and consequent problems such as salt contamination, the spreading of sewage and the fragmentation of coral reefs.

× **Landslides**: The intense rainfall could result in the saturation of hillsides making them more prone to landslides.

Figure 10: Damage caused by 'Superstorm' Sa in a neighbourhood near Brooklyn, New Yor

Case study: Hurricane Sandy, 2012

"Superstorm" Sandy first made landfall in Kingston, Jamaica on 24th October 2012, causing 70% of Jamaica's homes to lose electricity. The hurricane then moved in a northwards direction for five days until it reached New York on 29th October 2012. The speed at which the hurric. moved northwards was relatively fast at 25km/h. Hurricane Sandy was originally classed as a category 1 hurricane when it formed in the Caribbean Sea. However, as it travelled further north it gained latent heat from the warm waters of the Gulf Stream, causing the wind speed and storm surge to intensify into a category 3 hurricane.

In New York, 53 people were killed by the hurricane and the storm surge of 13.8 feet caused damage to the subway and a number of electric cables. The impacts of the hurricane were also felt on Long Island where 100,000 homes were damaged or destroyed and 8.1 million househ were without any power. Secondary effects included the closure of schools and universities for several weeks, the cancellation of 18,000 flig in and out of New York's busiest airports and incidences of looting for gasoline and heaters. The total damage caused by Hurricane Sandy wa close to $32 billion.

What are the characteristics of drought?

❖ A drought is a long-term and non-seasonal change in the 'average' conditions in the atmosphere (i.e. lack of rainfall). In arid areas, drought is seen as the intensification of dry conditions. There are four main types of drought: meteorological (long-term absence of rainfall), agricultural (deficiency of moisture in soil), hydrological (loss of stored water) and socio-economic (where human demand for water exceeds supply).

The spatial distribution and origin of drought

❖ Since the causes of drought are so varied, the distribution of droughts is very widespread. However, the areas most prone to droughts include: Southwest USA, Central America, Sub-Saharan Africa and East Africa, Southwest Europe and Southeast Asia. Both **MEDCs** (more economically developed countries) and **LEDCs** (less economically developed countries) can experience drought conditions.

❖ There are **three** main natural causes of drought conditions:

1. Variations and movements of the ITCZ
❖ The ITCZ moves from the northern hemisphere (June-August) to the southern hemisphere (December). However, on the long-term scale, the ITCZ may not always cover the same areas of land due to global climate change.

❖ As a result of this variation, areas that are ordinarily covered by the ITCZ do not receive their rainy season, leading to a period of dry conditions.

❖ **Example**: Sahel drought, 2000

2. El Niño
❖ An El Niño event occurs on average between every 2 and 10 years and can have a large impact on rainfall patterns.

❖ During December in a normal year, Australia and Indonesia have low pressure, whereas South America has high pressure. Thus, South America experiences dry conditions and Australia and Indonesia experience wet conditions. In June, the Walker Circulation restricts the westward movement of these warm waters, resulting in a period of low pressure in South America.

Figure 11: The Argentina drought of 2008 was caused by an El Nino event.

❖ However, during an ENSO (El Niño Southern Oscillation) event, the warm waters of the East Pacific Ocean move westwards due to the breakdown of the Walker Circulation that occurs every four or five years. As a result of this, South America will have a long period of high pressure and dry conditions, whereas Australia and Indonesia will experience a long period of low pressure and wet weather.

❖ **Example:** Argentina drought, 2008

3. Blocking anticyclones
❖ The upper westerly winds are very variable and they sometimes change track due to the occasional stretching of the air in the jet stream. This change in track leads to the upper level convergence of hot and cold air. This air piles up and eventually begins to descend as dry air. This air is deflected by the Coriolis force, creating an anti-cyclone.

❖ The formation of this anti-cyclone results in stable, calm and dry (high pressure) conditions, in which the amount of rainfall is decreased for a long period of time.

❖ Occasionally, anticyclone blocking can occur when the high pressure remains stationary for a while and 'blocks' the eastward moving depressions. Anticyclone blocking can maintain these dry conditions for the long-term (weeks/months).

❖ **Example**: UK drought, 1976

> **Exam hint!** It is important that you only mention the **three** causes of drought listed above in an exam question. Although other factors, such as high population density, deforestation and over-cultivation may increase our vulnerability to the impacts of a drought event, they do not contribute to the short-term changes in rainfall patterns that cause the dry conditions.

For case studies of **two** contrasting recent droughts refer to **Pages 23 and 24**.

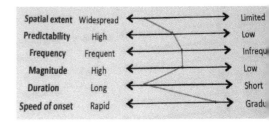

Figure 12: A typical drought hazard profile.

The characteristics of drought

1. **Spatial extent:** Although drought conditions (long-term shortage of rainfall) may only affect one community initially, the impacts of a drought event can become increasingly widespread over time. For example, when crop productivity in one area becomes reduced due to drought, emigration of farmers can increase the water-strain on foreign lands. This process is known as "spreading the impact".
2. **Predictability:** Drought is relatively easy to predict with new advanced satellite technology, which helps to show which communities may be experiencing loss of vegetation cover and lack of cloud cover due to a drought event.
3. **Frequency:** Drought events are relatively infrequent. This is because most droughts are caused by a period of several years with less than average rainfall. However, as global warming increases the variability of rainfall it is expected that the occurrence of drought will become more frequent.
4. **Magnitude:** Although most people perceive droughts as having a small impact, the truth is that the magnitude of drought is very high, but less visible. For example, droughts are "silent killers" as most of their impacts occur on a large time scale (e.g. death by famine) and have much less of an impact on infrastructure when compared to hazards, such as earthquakes and hurricanes.
5. **Duration:** The duration of a drought event may last from several months to several years (e.g. parts of the Sahel and northeast Africa have experienced drought since 2000). Although the amount of rainfall may increase during the middle of the event, the drought still occurs until the amount of water available for communities has been fully replenished.
6. **Speed of onset:** Drought is commonly known as having a gradual speed of onset and has subsequently been deemed a "creeping" hazard. Sometimes the period between the beginning of the drought and the first death caused by famine or water shortages can last for several years. Drought events have a slow speed of onset because most of the indirect impacts become increasingly severe over time. For example, as crop production continues to decrease, famine within the population will become increasingly severe.

Effects of drought

× **Famine:** Increased soil aridity reduces the agriculture potential of the soil, leading to a low crop yield. Famine can result in further problems, such as malnutrition, the spread of disease and inflation in the price of food products.

× **Biodiversity loss:** Owing to the lack of water, there is a large decline in the abundance and variety of flora and fauna. Lack of water is a particular issue for water-dependent species, such as goats. Biodiversity loss can have further repercussions, such as loss of tourism and loss of important ecosystem services, such as soil regulation.

Figure 13: Drought conditions can reduce the productivity of soils.

× **Conflict:** As water supplies begin to decrease, increased strain occurs, leading to violent competition over the limited remaining water resources.

× **Desertification:** Deficiency of moisture within a soil can make a soil more vulnerable to wind erosion, leading to soil degradation. Over a long period of time, desertification may result.

× **Economic loss:** As well as the occurrence of inflation, the primary producers of the affected crops (farmers) begin to lose their income. This is an especially large issue in LEDCs as their GDP (Gross Domestic Product) is generally highly dependent on the agricultural sector.

× **Salinization:** Due to the loss of vegetation from the lack of soil moisture, the amount of salt rising up from the ground increases, leading to salt-intrusion of the small supply of freshwater left available and the death of salt intolerant crops.

× **Forest fires:** Areas of woodland that are affected by drought are especially vulnerable to forest fires as sparks are more easily ignited in dry conditions.

What are the characteristics of technological hazards?

·y definition!

❖ A human-induced hazard is a man-made accident that originates from human error.
❖ There are many types of human-induced hazards, but the IB course requires you to focus on **technological hazards**. These are man-made accidents that originate from human error during the use of technology.
❖ Technological hazards usually involve the explosion or escape of hazardous material. For example:

  ➢ **Bhopal, 1984**: leak of methyl isocyanate into the surrounding environment. (See **Page 26**)
  ➢ **Chernobyl, 1986**: the explosion and leakage of nuclear radiation.
  ➢ **Deep-water Horizon, 2010**: the explosion and spillage of oil into the Gulf of Mexico.
  ➢ **Texas, 2013**: The leakage and consequent explosion of highly flammable ammonia. (See **Page 13**)

❖ Technological hazards can be exacerbated by environmental conditions, such as wind speed and the direction of a river's flow.

The spatial distribution of technological hazards

❖ Technological hazards are widely distributed around the industrialising twenty-first century world as they are associated with the misuse of technology.
❖ However, most technological hazard events have occurred in the developing world. This is as a result of human error or improper use of technological devices. Human error is likely to be greater in developing countries because safety regulations, such as emergency alarm systems within factories and chemical plants, are laxer. Furthermore, many TNCs (trans-national corporations) that own factories in NICs (newly industrialising countries) make many cost-saving changes (e.g. removing a refrigeration system) to save on labour costs, increasing the likelihood of a technological hazard.
❖ In addition, the infrastructure of LEDCs and NICs is less advanced than that of MEDCs, meaning that external response teams will be delayed, increasing a community's vulnerability to an event.
❖ Although it is easy to over-generalise the spatial distribution of any hazard type, it can be noted that, historically, a greater proportion of technological hazard events have occurred in either LEDCs or NICs. For example, between 1900 and 2011, 68% of all technological hazard events occurred in Africa and Asia.

Explaining the characteristics of technological hazards

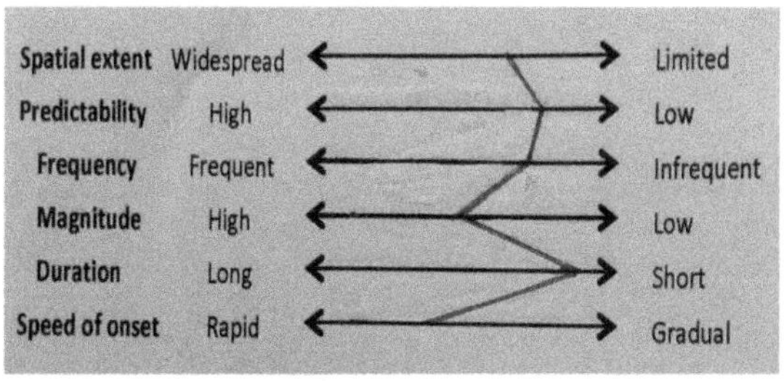

Figure 14: A typical hazard profile for technological hazards.

1. **Spatial extent:** Their extent is very limited as the explosion or leak of chemicals/gases only affects the settlements within close proximity to the hazard site. This is because, like earthquakes, the impacts of technological hazards obey distance-decay theory. Conversely, some technological hazards can have a very widespread spatial extent. For example, the radiation released from Chernobyl in 1986 was able to travel over 2400 km away to Britain. However, it should be noted that the radiation dose in Britain was at a much safer level than in Pripyat, Ukraine.

2. **Predictability:** Technological hazards are no more predictable than natural hazards. Although we know why, where and how human technology can fail, it is extremely difficult to predict when and under which circumstances human error will occur (e.g. Chernobyl and Bhopal were not predicted). On the other hand, we can assess the state and safety of technology on a regular basis and make precautionary modifications. However, communities or companies often ignore risk assessments due to personal hazard perceptions and economic reasons.

3. **Frequency:** On average, 65 technological hazards occur each year. They are much less frequent than natural hazards, such as earthquakes and hurricanes. The frequency of technological hazards is likely to decrease in the future, providing the effectiveness and ease of use of safety systems is improved. For example, following the Bhopal gas leak of 1984, many fertiliser companies in the USA spent millions of dollars on improving safety systems to ensure the frequency of these events was reduced.

4. **Magnitude:** The size of technological hazards is relatively large as technology can be extremely dangerous when improperly used. However, unlike natural hazards, the magnitude of these hazards

can be modified through emergency safety mechanisms. The magnitude of the event also depends on the extent of the human error.

5. **Duration:** The duration of the initial technological hazard (oil spill, nuclear accident) rarely exceeds a couple of hours. This is because response teams are able to modify the event by extinguishing a fire or destroying a toxic substance soon after the realisation of the hazard. However, some technological hazards have a much longer duration. For example, the pollution of the Yangtze River in China within the last decade can be seen as having more of an on-going environmental impact.

6. **Speed of onset:** The onset of the impacts of technological hazards is generally rapid. However, this does vary and is completely dependent on the hazard type. For example, an oil spill would have a more rapid onset due to its immediate impact on the environment (e.g. death of fish), whereas a nuclear explosion would obviously have some immediate impact, but the more severe and longer-lasting impacts are gradual and accumulative (e.g. development of leukaemia).

7. **Effects:** On the global scale, technological hazards are responsible for a tenth of the number of deaths caused by natural hazards. Although the immediate death tolls of these hazards are not very high, the effects of these events can be more long-term and can affect several successive generations. For example, the Bhopal gas leak that occurred 30 years ago is still affecting communities through post-traumatic stress, chemical-related diseases and birth defects. However, the effects of other technological hazards, such as the Deep-water Horizon oil spill, had much more of a short-term effect due to the effective clean-up strategies following the event.

Case study: Texas fertilizer plant explosion, 2013

On 17th April 2013, a fertilizer plant in the town of West, Texas exploded. This fertilizer plant was manufacturing chemical fertilizers using anhydrous ammonia, a highly explosive chemical. Although investigations into the origin of the explosion are still in progress, recent claims have suggested that the blast was triggered by a fire in a grain container on the site of the fertilizer plant. Following this, the fire is believed to have spread to the other 54,000 pounds of fertilizer, thus leading to a large explosion. Many have argued that the owners of the plant were responsible for the explosion as safety codes at the plant were not strictly enforced and they had failed to make the necessary adjustments following a damning safety report in 2006.

It has been estimated that 15 people were killed in the explosion, including three firefighters, who died attempting to put out the flames of the fire at the plant. Furthermore, over 160 people were injured by the event as a result of falling shrapnel and burning embers. The explosion was of such a high intensity that it triggered a magnitude 2.1 earthquake and eyewitness accounts have compared the explosion to a tornado or the dropping of a nuclear bomb. It is believed that between 50 and 75 properties were damaged by the blast with a further 20 properties being completely demolished, including one of the town's largest middle schools. The explosion of the fertilizer plant also had a short-term impact on local farmers who relied upon the use of fertilizers to increase their crop yield.

IB exam focus: Characteristics of hazards

(a.) Define the term, "Hazard". **[1 Mark]**
(b.) State **one** natural hazard type and describe its spatial distribution. **[3 Marks]**
(c.) Explain the causes of a **named** drought event. **[6 Marks]**
(d.) Compare and contrast the characteristics of **either** earthquakes **or** volcanic hazards and technological hazards. **[10 Marks]**

How should I approach this question?

(a.) You should use/learn the IB-preferred definition of "Hazard" (refer to **Page 3** for this definition).
(b.) Just state **one** natural hazard! There is no need to explain the distribution of the chosen hazard but you should just describe two detailed points about the distribution of the hazard type. For example, earthquakes occur at tectonic plate boundaries and along linear fault lines, hurricanes form in the tropics- but not close to the equator and drought can occur in areas of short-term high pressure. Answers describing the distribution of technological hazards will not be credited.
(c.) You must have a **named** example (e.g. The USA drought, 2012-2013) as this is a requirement of the question. The causes of the named drought should relate to the area affected by the drought. For example, variations in the movement of the ITCZ, blocking anticyclones and El Niño events. Each factor should be developed to secure 2 marks per factor.
(d.) This question requires two different accounts. One examining the characteristics of earthquakes/volcanic hazards and the other comparing these with technological hazards. Top answers will include case studies for both hazard types to access bands E and F. For each hazard, comment on (and give evidence of) spatial extent, predictability, frequency etc.

Section 2: Vulnerability

Why do people live in hazardous areas?

❖ It is estimated that 26% of the Earth's population live in hazardous areas. These areas might include coastal areas (at risk from hurricanes and tsunamis), slopes of volcanoes (volcanic hazards) and areas situated along fault-lines (earthquakes).

❖ People who live in hazardous areas are known as **vulnerable populations**. This is because they are more exposed to geophysical hazards and are therefore more susceptible to the impacts of hazard events. As the interaction between human and physical processes become more overlapped, the degree of risk posed by a hazard event increases. All things being equal, we would expect that the closer that populations live to the geophysical threat, the more vulnerable that population would be (See **Page 18** for the nature of risk).

❖ There are a number of reasons (both voluntary and forced) why people continue to live in hazardous areas, despite the high risk of loss of life, homelessness, injury and financial costs:

1. Hazards are unpredictable

❖ Although scientists can use some techniques to help predict hazard events, it still remains very difficult to predict exactly when, where, how often and how big a hazard event will be. As a result of this, vulnerable populations are unaware of the extent of risk posed by the hazard and continue to live in the hazardous area.

❖ For example, the densely populated area of Port au Prince was located on a blind-thrust fault. Subsequently, the populations living there have been unable to move away from the hazardous area.

2. Perceived benefits versus costs

❖ People living in hazardous areas subconsciously weigh up the benefits of living in a hazardous area against the costs posed by the occurrence of the hazard. If the benefits outweigh the costs then the populations are likely to persist in living in these locations.

❖ The return-period of the hazard event is also important in the costs versus benefits analysis process, as people are usually more willing to take more of a risk with low frequency, high magnitude events.

❖ Common benefits of living in hazardous areas include:

✓ **Fertile volcanic soils**: Although volcanic eruptions pose a large threat to the poor communities in poorly built housing of El Salvador, the rich fertile soils on the slopes of volcanoes provide large opportunities for coffee farming.

✓ **Tourism**: A large number of people go to visit hazardous areas due to the recreational services that these areas might provide and due to the fact that more non-hazardous areas (e.g. Amazon Rainforest) are being put under large human threats. For example, more people are visiting areas such as Japan (at risk from earthquakes/volcanic hazards) and Australia (at risk from drought) to visit Mount Fuji and Ayers Rock respectively.

Figure 15: Is coffee farming in El Salvador worth the risk?

✓ **Geothermal energy**: Although Iceland is at risk from volcanic hazards, geothermal energy provides a large proportion of its energy needs, meaning that the economy and sustainability of the country can develop.

3. Lack of alternatives

❖ People who live in hazardous areas have a number of social, economic and cultural ties, which prevent them from moving away.

❖ Social ties may include historical/cultural inertia- families and certain ethnic groups have always lived there, as it is central to their faith/tradition/culture. For example, the Masai in Kenya refused to move even in drought.

❖ Economic ties may relate to affordability. For example, the slum dwellers living in close proximity to the Union Carbide factory in Bhopal could not afford to live elsewhere and furthermore it provided employment opportunities.

4. Changing levels of risk

❖ When individuals initially move to live in an area, risk may not be present. However, as the frequency and intensity of hazards can vary in both time and space, the severity of risk posed to these individuals may increase due to physical or human processes, such as global warming and the introduction of new technology, such as fracking.

❖ Residents of South Sudan, for example, have become more at risk from drought as farmers have migrated from northern drought-stricken countries, putting increased pressure on both the land and water supplies.

What factors influence the vulnerability of a community?

Key definition!

❖ Vulnerability is the susceptibility of a community to hazards or the impacts of a hazard event.

❖ Vulnerability is a function of demographic and socio-economic factors, and of a community's preparedness and ability to deal with a hazard event when it occurs. The following factors influence the vulnerability of a community:

1. Wealth

❖ Developing countries are more vulnerable to hazards because they have less money to spend on hazard-safe infrastructure, emergency services and early warning systems. As a result of this, communities living in LEDCs are more susceptible to the impacts of a hazard event. For example, the residents of Burma were more vulnerable to the impacts of Hurricane Nargis because prediction was not published. As a result of this, people were unaware of what they were supposed to be preparing for.

❖ On the other hand, developed countries, such as USA, have better prediction and response systems and are thus less vulnerable to the impact of hazard events.

2. Preparedness

> **Home preparation**: By preparing a home for hazards (e.g. securing pictures and furniture to the wall before an earthquake or covering windows with shutters before a hurricane) will reduce the vulnerability to the impacts of a hazard.

> **Emergency services:** Once a hazard event is predicted, communities are less vulnerable to the risks posed as the emergency services are on "stand-by". For example, the residents of New York were given a three-day long warning of onset for Hurricane Sandy, meaning that emergency services were able to position themselves along the coastline and people were able to prepare as fully as possible by buying in extra groceries, boarding up windows of houses and constructing a line of sandbags along the coastline to act as a sea defence.

Figure 16: Preparation for Hurricane Sandy involved the use of sandbags to protect buildings from flooding.

> **Education:** Education is important in determining the vulnerability of a community in two respects. Firstly, a well-educated individual is more likely to become employed and earn a higher income. As a result of this, they would be able to live in a safer area and in a better quality house. Secondly, education on coping with hazards can increase preparedness and effectiveness of response of a community and thus decrease vulnerability. For example, Japan educates young people on what to do in an earthquake and the USA has an annual 'hurricane week' to raise awareness of hurricane safety.

3. Characteristics of a hazard event

❖ The characteristics of a hazard event are very important in determining the vulnerability of a community. For example, if the speed of onset of a hazard is rapid, the preparedness period is shortened and thus the population will be less prepared and more vulnerable to the hazard. In the case of an earthquake, if the number of after-shocks is high then the vulnerability of a community will be greater as the community will have less time to recover from the initial occurrence.

❖ Countries that suffer multiple hazards, such as Japan and the Philippines, are going to be more vulnerable to hazards than countries that experience a limited number of hazards, such as the UK, as they are susceptible to a greater frequency of hazard events (See **Page 26** for more details).

4. Ability to cope

➢ **Building design**: In many slum settlements, poor building design or inadequate building codes increase vulnerability. However, if communities such as Japan have hazard-proof buildings installed, then the vulnerability of the communities will be lower as they will be less at risk from the effects of the hazard event.

➢ **Aid:** Countries that receive foreign aid are more able to deal with the impacts of a hazard event. However, some governments refuse aid because it encourages dependency and negative stereotyping. For example, the Burmese military government initially refused international aid resulting in Burma's poor emergency services being put under strain and many areas of the community remaining vulnerable.

➢ **Insurance:** Although insurance for hazards is difficult to obtain, it can decrease vulnerability to hazards as it can help communities to cope better with the secondary effects. For example, house insurance may provide temporary accommodation after the event, reducing the effects of homelessness, such as the spread of disease.

Figure 17: An earthquake-proof building opened in Tokyo, Japan (April 2013).

Why are some sectors of a population more vulnerable than others?

❖ Demographic factors are also important in determining the vulnerability of a population. This is because certain sectors of a community are more susceptible to the effects of hazard events than others.

❖ The following sectors of a community are more vulnerable to hazards than others:

1. Age

❖ The two age groups most vulnerable to hazards are the young and the elderly.

❖ The young and the elderly are both vulnerable due to the fact that they are less mobile and hence less able to evacuate or protect themselves during a hazard event.

❖ In addition, both of these dependents are more susceptible to the secondary impacts of a hazard event. For example, the old are less able to deal with the extreme heat associated with drought events and the young are more likely to experience long-term psychological problems following a violent earthquake as they lack the understanding of what is happening.

❖ For example, the young were the sector of the population most vulnerable to the effects of the Chernobyl nuclear accident as they were more likely to develop long-term conditions from radiation exposure, such as cancers.

2. Gender

❖ Women are more vulnerable to hazards than men as they have poorer access to resources, work indoors and are more likely to be caring for a relative or family.

❖ For example, it was estimated that four times as many women died in the Asian tsunami of 2004 as the men were out at sea fishing whilst most women were indoors working or looking after the family. Consequently, they had less time to escape from the tsunami.

3. Income levels

❖ Sectors of a population with lower income levels will not be able to afford to live in hazard-proof housing in a safe location. Instead, they will be forced to live in particularly vulnerable areas, such as at the bottom of a slope or on flood plains.

❖ For example, nearly 2 million low earning Burmese lived on land less than 5 meters above sea level, making them more vulnerable to the storm surge of Hurricane Nargis.

❖ People with a lower income are also less likely to be able to afford insurance and thus would not be able to rebuild their homes following a hazard event.

4. Illness/blindness/mobility

❖ Sectors of the population who are more physically disadvantaged will be less likely to escape from a hazard event.

❖ People already suffering from illnesses, such as heart disease, may also be more vulnerable as their health condition might be aggravated by the stresses involved with evacuation and loss of possessions.

5. Race

❖ In some communities, racial issues may result in the marginalisation of ethnic minorities into areas that are more exposed to a hazard and are left to fend for themselves following a hazard event.

❖ For example, nearly half of the casualties of Hurricane Katrina in 2005 were low-income African Americans. This is because they lived in areas of poor infrastructure and were unable to move out of these vulnerable locations.

IB exam focus: Vulnerability

Study the adjacent map, which shows the areas worst affected by Cyclone Nargis in Burma.

(a.) Define the term, "vulnerability". **[1 Mark]**

(b.) Describe the pattern of vulnerability shown on the map. **[3 Marks]**

(c.) Explain why some sectors of a population are more vulnerable than others. **[6 Marks]**

(d.) "The vulnerability of a community is only determined by economic factors". Discuss this statement. **[10 Marks]**

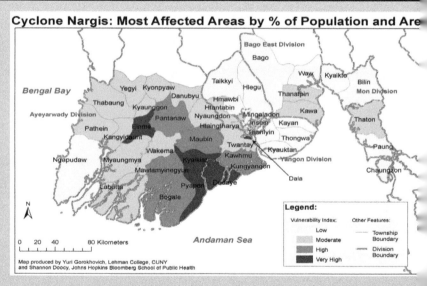

(Source: City University of New York)

How should I approach this question?

(a.) You should use/learn the IB-preferred definition of "vulnerability" (refer to **Page 15** for this definition).

(b.) When describing this map, you should describe the locations with the highest vulnerability (e.g. very high in south central and cent Burma) and the locations with the lowest vulnerability (e.g. low in the east and far west). In addition, you should try to identify the general trend shown on the map (e.g. the vulnerability to Cyclone Nargis is decreasing away from the centre) and note any exceptions. You will gain extra credit for using the scale to quantify your answer.

(c.) This question asks you to explain, but you will first need to identify, the vulnerable sectors of a population (e.g. the infirm, the poo the young, the elderly and women). You should then give reasons as to why these sectors are more vulnerable. These reasons sho focus on the sector's disadvantage in preparedness and ability to cope with a hazard.

(d.) To reach band D and above, there will need to be a discussion. This means that you must consider both sides to the argument give in the statement (i.e. "vulnerability is only determined by economic factors" and "vulnerability is determined by other factors"). To gain access to bands E and F, you should use supporting examples/case studies of the factors influencing vulnerability.

Section 3: Risk and risk assessment

ey definition! ❖ Risk is the probability of a hazard event causing harmful consequences (expected loses in terms of deaths, injuries, property damage, economy and environment).

What are the relationships between risk, predicted losses, probability and preparedness?

❖ There are **three** factors that combine to affect the degree of risk posed by a hazard. These are the predicted losses of a hazard event, the probability of a hazard event and the preparedness of the community. The relationships between these factors and risk are shown in **Figure 18**. The degree of risk is determined by the extent of overlap between the three factors.

❖ Risk is a very complex concept; although the probability of a hazard event cannot be reduced, the degree of risk can be lowered by increasing the community's preparedness and decreasing the predicted losses by modifying its vulnerability to hazard events.

❖ In order for scientists to perform a **risk assessment**, they must be able to determine the probability of the hazard, the predicted losses and the preparedness of a community.

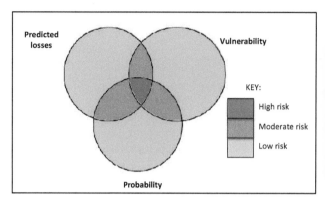

Figure 18: The relationships between risk, predicted losses, probability and preparedness.

1. Predicted losses

ey definition! ❖ The predicted losses of a hazard event are the likely impacts that a hazard event will have on a community in terms of deaths, injuries, property damage, economy and environment.

❖ If the predicted losses for a hazard event were high, the expected degree of risk posed by the hazard would be larger. This is because if the predicted losses are high then the probability of the actual hazard event causing harmful consequences will be higher. Scientists can make estimates for the predicted losses of a hazard event by using geographic information systems:

Geographic information systems (GIS) and hazard zone mapping

❖ GIS systems combine both the physical (e.g. spatial proximity, topography and magnitude) and human factors (e.g. vulnerability and preparedness) which influence the impacts of a hazard.

❖ They can be used to construct a hazard zone map, which shows the areas with the highest predicted losses and hence the areas most at risk.

❖ However, it can be quite difficult to accurately estimate the predicted losses of a hazard event using these systems because the factors influencing the impact of some events can suddenly change without warning. For example, Hurricane Katrina dissipated and reformed into a more violent storm of a higher magnitude resulting in the previous predicted loss estimates being surpassed by a significant margin.

2. Probability

definition! ❖ The probability of a hazard event is the likelihood of a hazard event actually occurring in terms of the magnitude of an event and the frequency of its occurrence.

❖ If the probability of a hazard event (and in particular, a high magnitude event) is high, the degree of risk posed by the hazard is greater. This is because a highly probable event (e.g. drought in the Sahel) is more likely to cause greater impacts than a less probable event (e.g. hurricanes in Britain), as it is more likely to actually occur and have an impact on a community.

Magnitude-frequency relationships

❖ Scientists can make estimates for both the predicted losses and probability of hazard events by examining historic trends of the hazard type and how likely a hazard of a certain magnitude is to occur. **Figure 19** shows that hazard events of high magnitude generally occur infrequently, whilst those of low magnitude generally occur more often.

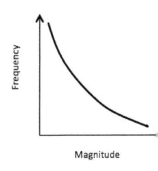

Figure 19: The relationship between the frequency and magnitude of hazard events.

- ❖ From studying the frequency of high magnitude events across time, scientists are able to deduce a **return period** for a hazard of a given magnitude. For example, in the UK the return period of a magnitude 5 earthquake is 20 years. However, this is just an idealised estimate and it is possible that more than one earthquake of this magnitude may occur within this period.
- ❖ However, the concept of magnitude-frequency is limited, as the magnitude of a hazard event is only measured by using a magnitude scale as opposed to an intensity scale. Subsequently, scientists are only able to deduce the size of an event and not the level of its impact. This is limiting because the predicted losses of a hazard event also depend on human factors, such as vulnerability and preparedness.

3. Preparedness

- ❖ Preparedness is the extent to which an individual or community is ready prior to a hazard event.
- ❖ Preparedness is influenced by both the characteristics of a hazard and socio-economic factors. For example, if the speed of onset is rapid, people are going to have less time to prepare themselves for the impacts of hazards. Also, if the magnitude of the hazard event is larger than anticipated, it is likely that all current preparation techniques will be rendered ineffective. For example, New Zealand building codes, which were built for a 50-year event, were not strong enough for the 2011 earthquake.
- ❖ If the probability of a hazard event is high, then it is likely that the community preparedness will also be high. However, preparedness might also depend on socio-economic factors.
- ❖ Socio-economic factors, which might influence the level of individual or community preparedness, include: education, communication, prediction, home preparation as well as emergency services and medical services.
- ❖ If a community or individual is less prepared for a hazard event, then the degree of risk posed by the hazard will be greater. This is because a less prepared community, such as that of Port au Prince, Haiti are more susceptible to the impacts of a hazard event and so the probability of the event causing harmful consequences is increased. The greater the intensity of the hazard event and the higher the vulnerability (low preparedness), the greater the risk posed by the hazard event.

Why do individuals and societies underestimate the probability of hazard events?

- ❖ Perceptions of hazard risks tend to be optimistic and unrealistic, leading individuals and societies to underestimate the probability of a hazard event occurring and lowering their preparedness for the hazard when it actually occurs. There are a number of reasons for this:
1. **Misplaced optimism:** Many people who lack previous exposure to hazards and value the benefits of living in hazardous areas often admit that they have never considered the potential risks. For example, following the Texas fertiliser explosion of 2013, many local residents admitted that they had perceived the plant as an agricultural necessity and never associated it with danger.
2. **Insufficient evidence for risk assessment:** Performing a risk assessment requires a strong ability to determine the predicted losses, probability and preparedness of a community. If a country does not have the technology or understanding to produce a hazard risk map then individuals and societies will be less aware of the risks and will thus underestimate them even if they are living in a hazardous area.
3. **Hazard regularity:** Many individuals and societies rely heavily upon the regularity of hazards when estimating the probability of hazard events. If a community were struck by a once in a hundred years hazard event, they would assume from the large return period that they were safe from the next hazard event i.e. "lightning never strikes twice". For example, following the 2010 New Zealand earthquake, communities perceived that another quake was unlikely. As a result, when the 2011 quake struck, the country was unprepared due to underestimating the probability of a similar magnitude event.
4. **"Technological fix":** Many societies often underestimate the probability of a hazard event because they believe that technology will be able to modify it and reduce its risk by, for example, diverting lava flows and using hazard-resistant buildings.
5. **Lack of awareness:** The risk posed by a hazard event may be underestimated as a result of lack of awareness. Through being unaware of the inherent risks associated with a certain hazard event, its true impact is less likely to be perceived. For example, lack of communication and media attention in

Burma before and during Hurricane Nargis meant that individuals were left uninformed of the risks posed by the hazard event.

6. **The risk of hazards versus other concerns:** In communities that are facing other issues, such as famine, economic decline, disease and conflict, individuals may underestimate the probability of a hazard event, as they will be pre-occupied with other concerns. Other societies may believe that "things can't get any worse". For example, the people of Haiti were facing the issues of poverty and crime prior to the earthquake, which diverted their attention away from the likelihood of a devastating natural hazard.

What factors determine an individual's perception of the risk posed by hazards?

•y definition!

❖ Risk perception is our individual estimation of the risks posed by a hazard event.
❖ The degree of risk perceived is nearly always lower than the actual risk posed by a hazard event. This is because we receive, filter and distort available information from scientists and even our own experiences. There are many factors that influence the way in which we filter and distort information:

1. **Past experience of hazard events:** If someone has experienced a traumatic hazard event, or knows someone badly injured or killed in one, they are more likely to fear it. However, an infant or tourist in a hazardous area would be unlikely to perceive a risk until its actual occurrence.

2. **Personality:** A person who is a risk-taker is more likely to perceive a hazard event as less of a threat than one who is risk-adverse and thus more likely to fear and make an attempt to escape any risks of a hazard.

3. **Level of education:** Education is closely linked to increasing the awareness of hazards. For example, earthquake drills in Japanese primary schools increases children's awareness of safety and hence enhances their perception of earthquake risks. However, being uneducated about the dangers of hazards results in less certainty about the potential risks and consequently more fear of the hazard.

4. **Economic status:** People able to afford to live in strong and stable houses will have less fear of the risks posed by a hazard event, as they will have more confidence in their ability to cope with it. However, those living in informal settlements are more likely to be unaware of the risks posed by the hazard and more susceptible to its impacts, thus increasing the level of fear associated with it.

5. **Religion:** Individuals with a strong religious belief perceive hazards to be naturally occurring "acts of God" that should not be feared, but accepted as a part of every day life.

What methods can be used to predict the probability and impacts of hazard events?

❖ Hazard prediction is very important in terms of reducing the impacts of a hazard event. If we know when the hazard event will occur in time and where it will occur in space, it is more likely that communities at risk will be given a longer warning of onset, thus giving them more time to prepare, alert emergency services or leave evacuate the hazardous area.
❖ Subsequently, impacts of hazard events such as loss of life, injury and property damage should be minimised as the vulnerability of the community is reduced.
❖ When predicting hazard events in time and space, scientists use both **probabilistic** and **deterministic** methods.
❖ Probabilistic methods are used to estimate how probable a hazard is in time and space from using information, such as historic trends of the hazards, the return period and scientific knowledge and understanding of the spatial distribution of the hazard.
❖ Deterministic methods are often used in the short-term period before the occurrence of a hazard. These methods aim to deduce the magnitude of the hazard event as well as a more accurate time-scale of its occurrence. Deterministic methods usually involve the direct monitoring of suspected hazardous areas (e.g. the use of a strain meter along a fault line or buoys in tropical areas).

Earthquake prediction

Probabilistic methods

1. **Plate tectonic theory:** From a detailed knowledge of tectonic theory, scientists are able to predict the areas most prone to earthquakes as being those situated along the edge of a tectonic plate. However, our geological knowledge is limited. For example, scientists are not completely certain why earthquakes, such as those that occur in the British Isles, do not occur along or near plate boundaries, making these earthquakes quite unpredictable. Furthermore, some earthquakes occur along a blind-thrust fault (a thrust fault "buried" underneath the upper crust that cannot be seen from the surface) giving scientists no chance to predict the probability of the earthquake (e.g. New Zealand, 2011).
2. **Magnitude-frequency relationships:** Scientists can use magnitude-frequency relationships to deduce the return-period of an earthquake of a given magnitude. However, the return period of an earthquake only gives scientists a rough estimate as to how often the event will occur as many see earthquakes as having more of a random temporal spacing.
3. **Seismic "gap" theory:** Seismic "gap" theory relates the probability of an earthquake to the frequency and magnitude of previous earthquakes along the same fault line. For example, if a fault line experiences lots of frequent earthquakes of a low magnitude, then the probability of a high magnitude earthquake would be low as the fault line has a small gap without seismic activity. By contrast, the North Anatolian fault line in Turkey has not experienced significant earthquake activity for over four years, meaning that scientists anticipate that the next earthquake to occur along this fault line will be of magnitude 7.0 or higher.

Deterministic methods

1. **Foreshocks:** Foreshocks are believed to be small seismic waves that ripple out from the focus of an earthquake in the hours before the occurrence of a high magnitude earthquake. Scientists can measure the occurrence of foreshocks along a fault line by using **seismographs** and **tilt-meters**, which detect minor earth movements. For example, foreshocks were detected 45 minutes prior to the Turkey earthquake of 1999, giving the community a lengthened preparation period. However, one limitation of this method is that foreshocks are only usually recorded in areas of unconsolidated rock.
2. **Animal behaviour:** Unusual animal behaviour is often associated with earthquake activity. Scientists suggest than certain animals can sense the chemical changes of groundwater associated with stressed rocks in the earth's crust. For example, a month before the Haicheng earthquake (1975) snakes were seen coming out of hibernation prematurely.
3. **Fault monitoring:** There are many different techniques that scientists can use along a pre-existing fault line to predict an earthquake. However, many of the recorded changes happen just minutes before the initial shaking meaning that these techniques do not act as an effective early warning system. Good examples of monitoring include the use of a **strain meter** to monitor changes in rock stress levels, the use of **radon gas counters**, which measure the dissolved gas from the strained rocks in the groundwater and the use of **lasers**, which can detect small amounts of movement along the fault line.

Case study: Predicting the L'Aquila earthquake, 2009

On 6th April 2009, a magnitude 6.3 earthquake hit the central Italian city of L'Aquila, which is situated along one of the country's largest fault lines. The earthquake killed 309 people, caused thousands of buildings to collapse and an estimated $15 billion worth of damage. Recent research suggests that both scientists and wildlife predicted this earthquake to a certain extent. However, prediction is useless when unused or unnoticed.

As L'Aquila is situated along a major pre-existing fault-line with a long seismic gap, seismologists were already monitoring the area. Following the earthquake, six of these seismologists were arrested for manslaughter as in the months before its occurrence, they had recorded a large number of foreshocks, but were uncertain of the risk posed by the upcoming event and thus gave incomplete and contradictory information of the potential dangers to the local community. This occurrence reaffirmed the fact that it is difficult to predict exactly where and when an earthquake and the dangers of that scientific uncertainty can cause. It has also been suggested that unusual animal behaviour acted as a warning sign that people chose to ignore in the days before the devastating earthquake. For example, colonies of toads from L'Aquila abandoned their mating site 3 days prior to the earthquake (as they are sensitive to changes in environmental chemistry) and returned after the majority of the after-shocks had stopped.

Hurricane prediction

Probabilistic methods

Figure 20: Satellite monitoring can be used to track hurricanes.

1. **Knowledge of the conditions required:** Scientists know that hurricanes can only form in oceanic areas with a sea temperature of 26.5°C, a depth of 70 meters and a strong Coriolis force. Hence, they will be able to focus on implementing monitoring techniques in areas with a high hurricane risk, such as the Caribbean Sea.

2. **Computer models:** In 2007, the National Hurricane Center (NHC) introduced a high-resolution NOAA (National Oceanic and Atmospheric Administration) hurricane research and forecasting model. These computer models are able to produce probability forecasts for hurricanes. Hurricane forecast models use equipment, such as aircraft, buoys and weather radars, to record pieces of useful meteorological data, such as wind speed, humidity, temperature and cloud cover. Along with this data, knowledge of occasional climatic variations, such as El Nino and La Nina, is used to produce a probability forecast, which lets us know the frequency and likely distribution of hurricane events in a particular year.

Deterministic methods

❖ **Predicting the hurricane's trajectory:** Scientists can attempt to predict the path of a formed hurricane by using satellite-imaging technology. The orbiting satellites take regular images and video footage of a hurricane to determine the direction and speed of the storm's movement. However, satellite monitoring is limited as it still fails to predict the potential impact of a hurricane when it makes landfall as the intensity of a storm can change rapidly and with little warning. For example, Hurricane Katrina, which was initially a low-magnitude hurricane, dissipated in the Gulf of Mexico. However, less than a day later the hurricane reformed and intensified to a category 5 hurricane and thus the population of New Orleans was unprepared for impact of the event.

Case study: Predicting Hurricane Nargis, 2008

Hurricane Nargis struck the town of Labutta, Burma on 2nd May 2008. Burma is situated along the coast of the Bay of Bengal. The Hurricane killed over 134,000 people and caused widespread damage to Burma's infrastructure, resulting in millions becoming homeless. It is estimated that over 65% of Burma's paddy fields were flooded by the storm surge, causing loss of livelihood in the agricultural sector and a significant inflation in food prices. Total damages caused by Hurricane Nargis are estimated to have been in the region of $10 billion. Hurricane Nargis had a relatively long speed of onset as the hurricane was travelling northwards from the Indian Ocean at a speed of only 11 km/hr. Since the storm was moving slowly, prediction of the hurricane should have been much more efficient.

The Indian Meteorological Department knew about the hurricane over 48 hours before the formation reached the coastline and alerted the Burmese military government. However, the military government failed to issue a timely warning to the vulnerable population situated in the path of the storm. Although some warnings were given on the television, prediction seemed very vague to the general public as it lacked details on the intensity and potential threat of the storm. As a result of this, people were unaware of what they were supposed to be preparing for. The Hurricane's intensity was very unpredictable as it started off as a category 1 storm and rapidly intensified into a category 4 hurricane without warning.

IB exam focus: Risk and risk assessment

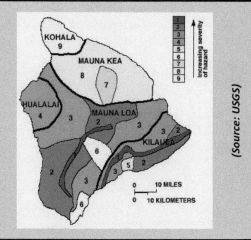

(Source: USGS)

Study this hazard zone map, which classifies different regions of Hawaii by their severity of risk of lava flows.

(a.) Describe the pattern of high-risk severity shown in the map. **[2 Marks]**
(b.) State **two** reasons why individuals and communities may underestimate the risk of lava flows in Hawaii. **[2 Marks]**
(c.) Explain the methods used to make predictions of the probability of **either** earthquakes **or** volcanic hazards occurring. **[6 Marks]**
(d.) Discuss the factors that combine to influence the degree of risk posed by a hazard. **[10 Marks]**

How should I approach this question?

(a.) This question is asking you to describe how high risk severity is spatially distributed. You will get no credit for describing anything other than the location of these areas.

(b.) Reasons for underestimating the risk of the lava flows could include; faith in lava diversion, lack of hazard event experience and the character of individuals (e.g. are they a risk-taker?) It is important that you only "state" reasons as going into unnecessary detail here could cost you time in the exam room and you will not gain extra credit.

(c.) The command term, "explain" requires you to give a detailed account of how **either** earthquakes **or** volcanic hazards can be predicted. You are recommended to mention at least three detailed prediction methods. For volcanic hazards, you should explain how scientists monitor geophysical changes, such as the release of sulphur emissions and changes in the physical appearance of the volcano. If you choose to answer this question on earthquakes, you could mention how we can use both our geological understanding of plate tectonics and monitoring techniques along the fault line, such as using a tilt-meter and measuring changes in the level of ground water and observation of unusual animal behaviour.

(d.) The factors that you should discuss in this question are predicted losses, probability and preparedness. For each of these three factors, you should explain how each are determined and how each contributes to the degree of risk posed by a hazard event. Do not simply make statements, such as "low preparedness results in high risk", but explain the relationship by saying that low preparedness makes a population more vulnerable to a hazard event and hence increases the risk associated with that event. To reach bands E and above, it is important that you refer to particular examples in your answer. For example, you could say that the densely populated city of Port au Prince in Haiti was more at risk from the 2010 earthquake due to the fact that national preparedness was low and too many people lived in sub-standard housing.

Section 4: Disasters

What is the difference between a hazard event and a disaster?

Key definition!

❖ A **hazard event** is the occurrence or realisation of a hazard, the effects of which change demographic, economic and/or environmental conditions. On the other hand, a **disaster** is a major hazard event, which causes widespread disruption to a community or region that the affected community is unable to deal with adequately without **outside help**.

❖ The distinction between a hazard event and a disaster is not always objective. This is because each community will have different criteria as to what counts as "widespread destruction". For example, one community might see the collapse of a thousand buildings as a disaster, whereas another might see the loss of a single human life as a disaster. It is important to remember that both these terms are relative and vary with the perceptions of individuals.

❖ Since the distinction between hazard events and disasters is not always clear-cut, two case studies of drought are shown below. One was considered a hazard event and the other a disaster. Can you distinguish between them?

Case study: Drought in the USA, 2012-2013

The USA drought began in spring 2012. The drought was caused by reduced snowfall in the northern parts of the USA. Subsequently, little water was able to percolate downwards into the soil. In addition, the snow that was falling was not melting due to cold temperatures. Reduced snowfall in the winter of 2011/2012 was caused by the North Atlantic Oscillation, which reduced the frequency of winter storms. A result of this decline in melt water, the rate of evapotranspiration decreased leading to reduced rainfall in the spring. This reduced rainfall gave rise to a meteorological drought. By June 2013, the eastern states had become drought-free.

Over 80 people died as a result of the extreme heat conditions. Furthermore, deficiency of moisture in soils has had a large impact on the yi of crops, such as corn and soybean. It is estimated that the 2012 corn harvest was 75% lower than average. As a result of this, an inflation o 4% on food prices occurred between 2012 and 2013. The drought also resulted in the increased occurrence of forest fires. For example, the Waldo Canyon fire in Colorado on 23rd June 2012 killed 2 people and destroyed 346 buildings. Overall, it is estimated that this drought has been the most expensive drought in the USA since 1956, as it is believed that damage costs will exceed $150 billion.

Despite the severity of this drought, the USA has been able to cope without the need for outside assistance. The main responses to the drought have included government-enforced restriction of water in the worst affected states. These restrictions have included hosepipe an sprinkler bans and have been effective in preserving the little remaining water resources. In August 2012, 33 states were designated as bein drought-disaster areas and were given $170 million worth of food imports, such as pork and chicken, from the USDA (US Department of Agriculture).

Case study: East African drought, 2011

The drought experienced by Eastern Africa occurred from July 2011 to August 2012 was the worst drought to occur in this area for over 60 years. This drought had a very large spatial extent, affecting four countries within the Horn of Africa: Ethiopia, Somalia, Kenya and Djibouti. The East African drought of 2011 occurred as a result of an unusually strong La Niña event in 2010, which caused rainfall in Kenya and Somalia to fail for two years. Lack of rainfall in these countries resulted in a deficiency of moisture in the soil, making agriculture more difficult. Some scientists believed that this unusually strong La Niña event was a consequence of global climate change, which increases windy conditions that blow moisture-containing clouds away in equatorial countries.

Figure 21: Dadaab refugee camp, Kenya.

It is estimated that between 50,000 and 260,000 people died as a result of the 2011 East African drought, most of which were caused by malnutrition and dehydration. The drought conditions resulted in a 30% decrease in rainfall in the affected regions and consequently resulted in large-scale crop failure due to the deficiency of moisture in the soil. As well as this, 20,000 cattle were killed by the extreme heat conditions. As a result of lowered crop yields and death of livestock, food shortages occurred within the Horn of Africa. The Famine caused by the drought led to increased malnutrition in Eastern Africa. For example, 30% of children in Somalia were suffering from malnutrition.

Due to lack of water and food resources in surrounding countries, the communities of East Africa were heavily dependent on outside help. Shortly following the occurrence of the drought, it was estimated that over $1.3 billion of aid was given by international non-governmental organizations, such as Oxfam and Save the Children. This aid was distributed in the form of bottled drinking water and non-perishable emergency food supplies. Another form of international food aid was "Doctors Without Borders", in which malnourished refugees were treated in a therapeutic feeding centre.

What methods can be used to quantify the spatial extent and intensity of disasters?

Spatial extent

- ❖ The spatial extent of a disaster can be quantified by using interactive mapping. Some scales, such as the Palmer Drought Severity Index (PDSI), can be used relatively easily to produce a disaster map showing the extent of drought in areas from +4 (normal conditions) to -4 (severe drought).
- ❖ For disasters resulting from earthquake or hurricane events, we can measure the spatial extent by using open source software, such as Ushahidi™. These software companies can collect disaster information from people affected via mobile phones, uploaded photos and social networks. This information can be instantly plotted on a regional or national scale map to show the size of the area affected by the disaster.
- ❖ For example, during the 2010 earthquake in Haiti, hundreds of earthquake-related incidents were recorded using this software and the affected areas were plotted on a map of the island to let aid workers and rescue services know where to focus their efforts.

Intensity

- ❖ We can quantify the intensity of disasters by using magnitude scales and/or intensity scales.
- ❖ Magnitude scales (e.g. the Moment-Magnitude scale) are those that measure the size of the geophysical event that results in the disaster, whereas an intensity scale (e.g. the Mercalli scale) assesses the overall impact of a disaster in terms of loss of life, property damage and injury.
- ❖ Magnitude scales are often favoured over intensity scales because the intensity of a disaster can be subjective. The Saffir-Simpson scale combines both the magnitude and the intensity of a hurricane.

1. The Moment-Magnitude scale (MW)

- ❖ Although the Richter scale is often still referred to in the media, seismologists have used the Moment-Magnitude scale to measure the magnitude of earthquakes since the late 1970s.
- ❖ Unlike the Richter scale, which only measured the energy of the seismic waves released during an earthquake, the new scale measures the seismic moment of the earthquake. The seismic moment is equal to the rigidity of the earth multiplied by the average amount of slip on the fault and the size of the area that slipped.

- The Moment-Magnitude scale is a logarithmic scale, meaning that each ascending unit is 10 times more powerful than the previous unit. For example, the 2011 Japan earthquake (MW 9.0) was over 100 times stronger than the 2010 Haiti quake (MW 7.0).
- It is difficult to place a critical value on the Moment-Magnitude scale because the amount of damage caused by an earthquake also depends on human and other physical factors. However, we usually consider an earthquake above magnitude 4.0 on the scale to be a potential threat and a quake above the magnitude of 7.0 to be disastrous in most cases.

2. The Saffir-Simpson Hurricane scale (SSHS)

- Hurricanes are measured using the Saffir-Simpson scale, which consists of four categories based on their central pressure, wind speed, height of storm surge created and their damage level.
- There is no maximum or minimum limit for each of the categories, but it has to be at least 74 mph to be classified as a hurricane.
- The critical value on this scale is category 3. Hurricanes of category 3 and above are considered to be disastrous. Recent examples of these include "Superstorm" Sandy, Hurricane Katrina and Hurricane Nargis.
- However, it should be noted that category 5 storms don't always cause the most damage, as the amount of damage caused is also highly dependent on human factors such as population density and community preparedness.

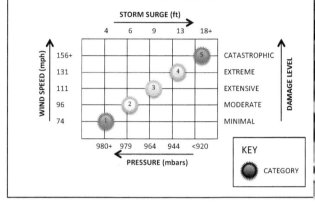

Figure 22: The Saffir Simpson Hurricane scale.

What were the causes and consequences of a disaster resulting from a natural hazard?

Case study: Tohoku earthquake and tsunami, 2011

Causes

The Tohoku earthquake occurred 130 km off the coast of Sendai, Japan on 11th March 2011 at 2.46 PM and measured a catastrophic 9.0 on the Moment-Magnitude scale. It was the largest Earthquake to be recorded in Japan in living memory and the third most powerful earthquake to occur in recorded history. Japan is well known for its frequent seismic activity as it is situated along a destructive plate boundary where the Pacific plate sinks underneath the Eurasian plate. However, this subduction process is not smooth and friction caused the Pacific plate to stick and to become 'locked'. As a result of this, pressure began to build up along this plate boundary and suddenly released as an immensely strong earthquake. The earthquake had a shallow focus of 30 km and the initial ground shaking lasted for approximately six minutes.

Consequences

It is estimated that over 15,400 people were killed as a result of the Tohoku disaster and over 2 million people were left homeless. In addition, power lines, gas and water services were severely disrupted as the earthquake ruptured underground cables and pipes. However, the majority of the impacts occurred as a result of the tsunami that was generated by the earthquake and hit a large area of the eastern coast of Japan. This tsunami was 10 m in height and travelled at a colossal speed of 800 Kilometres per hour. The tsunami contaminated water supplies, destroyed homes and caused widespread disruption to road and rail services, leading to a delay in rescue responses. Furthermore, the tsunami caused the cooling system at Fukishima Daiichi nuclear power plant to fail and subsequently caused radioactive materials to escape, causing local radioactivity levels to increase by up to 40,000 times.

Figure 23: The 2011 quake was the largest earthquake to hit Japan in recorded history.

The total cost of both the earthquake and the tsunami was estimated at $185 billion. Responses to the disaster were very efficient and within less than an hour of the earthquake and tsunami, the Japanese Red Cross sent 230 teams to the worst affected areas to provide medical and emotional support to the victims. Within the first month of the disaster, many NGOs responded to the disaster. For example, Shelter Box sent 1,500 boxes of emergency aid to the victims.

Although 75% of Japan's buildings are earthquake-proof, the sheer intensity of the earthquake was much higher than anticipated on the eastern coast of the Island and therefore exceeded the capacity of the buildings. Furthermore, Japan has an ageing population, meaning that less people were able to evacuate in the short amount of time given. As the epicentre of this earthquake was in close proximity to Sendai, a largely populated city, the damage caused was far greater as there were more potential fatalities.

What were the causes and consequences of a technological disaster?

Case study: Union Carbide; Bhopal gas leak, 1984

Causes

As a result of rapid population growth in India during the late 20th century, the demand for agriculture was increasing. In order to produce a greater yield, the large-scale manufacturing of agricultural pesticides and fertilizers began. In the 1970s, a USA trans-national company, Union Carbide, established a pesticide factory in Bhopal. However, on 2nd December 1984, water entered a storage tank containing methyl isocyanate (MIC). This water reacted with the MIC in an exothermic reaction causing the temperature, and hence pressure, in the storage tank to increase. As a result, gas began to leak out into the surrounding environment. It is estimated that a total of about 30 metric tons of gas was released over a 60-minute period.

Consequences

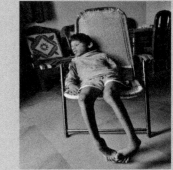

Figure 24: The gas leak caused many unborn children to be born with cerebral palsy.

The chemicals present in the gas cloud were able to attack internal organs preventing oxygen from entering the blood stream and over 2200 people died immediately as a result of this poisonous gas. However, those who were not killed by the gas experienced severe symptoms, such as coughing, vomiting and suffocation. The Bhopal disaster also had a large environmental impact as the gases caused visible damage to surrounding trees, causing them to lose their leaves within a few days. It has recently been found that many pregnant women exposed to the gas gave birth to children with congenital defects and low birth weight. Furthermore, studies have shown that the incidence of certain medical conditions has become more frequent amongst the population affected by the gas leak, such as miscarriages and lung cancer. Responses to this technological disaster were very limited and solely focussed on the distribution of financial aid to the affected populations. However, the amount of aid given was insufficient. For example, it was estimated that the Indian government only distributed £4 worth of aid for each family affected by the gas leak. The Union Carbide factory was closed on 16th December 1984 and any remaining MIC was converted into a safe form.

The impact of this technological disaster was largely due to the fact that a slum settlement, containing over 12,000 people, was located just outside the gates of the factory, hence providing a large vulnerable population. Owing to India's poor economic status, Bhopal had very poor emergency services and healthcare, leading to increased vulnerability to gas-related diseases. As a result of this high population density, more people were vulnerable to inhaling the gas and medical services were sparse within the slum settlement. Since Union Carbide was a money-maximising trans-national corporation, little money was spent on safety checks and in June 1984, the company decided to close down the factory's refrigeration system to cut costs. However, this decision was to increase its vulnerability, as it led to increased temperature and pressure within the storage container, giving rise to the deadly leak.

How do the intensity and impacts of disasters vary in space and change over time?

❖ Disasters are a complex and non-static phenomenon. The intensity and impacts of disasters vary in space and changes across time:

The varying impacts of disasters in space

❖ **LEDC vs. MEDC**: Developed countries experience more economic damage, whereas less developed countries experience more social losses, such as loss of life and injury. This is because people in MEDCs are more prepared and thus more able to escape the hazardous area, whereas in LEDCs they are less prepared and thus experience more social losses. However, there are some exceptions to this trend. For example, mega-disasters, such as the Japan earthquake of 2011, can bring both large economic and social losses to developed regions.

❖ **Proximity to hazardous area:** The intensity and impacts of disasters decrease as the distance away from the hazardous area increases by a process called distance-decay.

❖ **Physical environment:** The intensity and impacts of disasters can also vary dependent on the physical environment in which the disaster occurs. For example, the intensity of an earthquake would be larger if it occurred in an area of unconsolidated rock due to the fact that liquefaction would occur and cause damage to buildings. The impacts of a disaster would be reduced if a hurricane occurred in a rural area, due to the fact that there would be fewer buildings to damage.

The varying impacts of disasters over time

❖ The number of people affected by disasters has changed over time due to the fact that more people are putting themselves at risk by living in vulnerable hazardous areas.

❖ The number of people living in hazardous areas is increasing due to population growth and the declining supply of housing.

❖ The number of disaster-related deaths varies from year to year. For example, the disaster death toll in 2005 was over 90,000 people, whereas in 2010 there were over 300,000 disaster-related deaths. MEDCs have seen a decreased number of disaster-related deaths due to improvements in preparation and prediction techniques. However, LEDCs have seen little drop in the number of disaster-related deaths, as many of these countries lack the capital to make the necessary adjustments.

❖ However, the amount of economic damage caused by disasters on the global scale is increasing. For example, the damage caused by the Kobe earthquake of 1995 was $102.5 billion, whereas total damage costs for the Japan earthquake of 2011 was over $185 billion.

❖ The economic loss of disasters is mainly increasing in developed countries, such as the USA and Japan. This is because people have more valuable possessions over time, the value of infrastructure and property has increased over time and more people are now living in hazardous areas.

❖ In recent years, it has been suggested that increased human activity has affected both the frequency and intensity of disasters. For example:

➢ **Earthquakes** can be triggered by human processes, such as mining and fracking. Minor earthquakes in the UK have been believed to be caused by fracking.

➢ **Hurricanes** and **droughts** can become more intense as the frequency of short-term changes in the atmosphere, such as El Niño and La Niña, have increased due to global climate change.

➢ **Technological disasters** have become more frequent with the industrialisation of developing countries and the intensity of these disasters increases as human ingenuity leads to the use of more hazardous technology, such as nuclear power.

IB exam focus: Disasters

Study the proportional circles map, which shows global disaster damages from 2007.

(a.) Explain the pattern shown on the map. **[4 Marks]**

(b.) Outline the consequences of a **recent** technological disaster. **[6 Marks]**

(c.) With reference to examples, discuss the ways in which the intensity and impacts of disasters can change over time. **[10 Marks]**

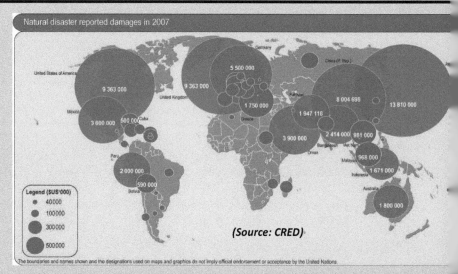

(Source: CRED)

How do I approach this question?

(a.) This question requires you to describe the map and then give reasons for the pattern of global disaster damages. You should start b stating where the most disaster damage occurred in 2007. To gain extra credit, you should use the map key to help quantify your answer. You should then describe where the least damage was found. Reasons for higher damages in MEDCs could include larger c more expensive buildings and infrastructure and high population densities.

(b.) The command term "outline" asks you to give a brief account. You should outline both the short-term and long-term consequences your chosen technological disaster. Although the question asks you to write about a recent technological disaster, examples, such a Bhopal and Chernobyl are valid so long as you stress that the impacts of the disasters are on-going and still affecting communities today. You should not use examples such as the Texas fertiliser explosion and Deepwater Horizon since no outside help was require thus they were hazard events as opposed to disasters.

(c.) In your answer, you should offer a balanced and exemplified review. You should consider how the economic costs of disasters have increased over time and how the social costs have decreased as preparedness has increased. However, you should mention that m disasters, such as the Japan earthquake of 2011, still have a catastrophic impact on a country's society and economy.

Section 5: Adjustments and responses to hazards and disasters

How useful is risk assessment in selecting adjustment and response strategies for a hazard?

y definition!

- ❖ Risk assessment is the process of establishing that a hazardous event of a particular magnitude will occur within a given period. It also involves making estimates of the impact of the hazardous event by taking into account the location of buildings, facilities and emergency services.
- ❖ Risk assessment is useful in deciding the strategies of adjustment and response to a hazard because it may help identify particular places and people for whom the risk of the hazard is greater. For example, if we know that the southern coast of the USA is at high risk from hurricanes, we can adjust the impacts of the storm surge by maintaining a belt of mangroves or building strong flood barriers along the coastline.
- ❖ If it is known that a particular area is at risk, such as an informal settlement or re-claimed land, governments are able to concentrate and spend more money on the emergency services in that location. By doing this, the response to the actual hazard event will be more efficient and will help to reduce its impact. In addition, governments can encourage community preparedness in these areas by increasing their awareness of the intensity and risk of the upcoming disaster.
- ❖ Risk assessments can also be useful in determining the strength of the adjustments that will be required. For example, seismologists know that Japan is at risk from experiencing a magnitude 7.0 earthquake between every 5 to 10 years, meaning that the country can retrofit its buildings to cope with these events. However, risk assessments prove as good as useless when making adjustments for freak mega-disasters, such as the 2011 earthquake.

What attempts can be made to reduce vulnerability to the impacts of hazard events?

- ❖ The susceptibility of a community to the impacts of hazard events can be reduced by spreading the risk and by land use planning.

1. Spreading the risk

Aid

- ❖ Disaster aid is the distribution of financial or practical support to help with rescue, rehabilitation and reconstruction responses, the cost of which is shared throughout the tax-paying population. However, it can also be given by NGOs.

Figure 25: The distribution of aid following the 1994 Northridge earthquake.

- ❖ This aid can be given at the national or local level both during and after the disaster. During the disaster, aid such as medical teams, food and shelter are provided and after a disaster, aid can be given to help invest in education and prediction systems to increase preparedness for future hazard events.
- ❖ Emergency aid is important in reducing vulnerability as it increases a nation's ability to cope and recover following a major hazard event or disaster. It is likely that without this aid the affected communities would have to bear the losses.
- ❖ For example, aid was very useful in reducing vulnerability to the impacts of the Indian Ocean tsunami of 2004 as NGOs, such as Unicef, provided temporary schools for 500,000 children, which aimed to return their lives back to normality and reduce their vulnerability to long-term trauma.

Insurance

- ❖ Insurance is a very costly way to spread the risk of a disaster since insurance companies charge according to the degree of risk posed by a hazard and in areas of very high risk insurance may not be available. As a result of this, insurance is a strategy used mainly in MEDCs. However, due to lack of risk perception not many people take out insurance for hazards. For example, only 10% of California's population are insured against disaster damages even though over 10,000 earthquakes occur in California each year.

- ❖ Insurance encourages increased preparedness due to the fact that some companies require building and preparing homes in accordance with a long list of safety codes. The more prepared for the hazard event a community is, the less vulnerable it will be to its impacts.
- ❖ Home insurance may provide temporary accommodation after the event meaning less vulnerability to the effects of homelessness such as the spread of disease. In addition, the cost of loss of possessions is often covered and therefore those affected are less vulnerable to the economic cost of disasters.
- ❖ However, insurance for hazards, such as hurricanes, is becoming more difficult to obtain due to the increased frequency and intensity of these events.

2. **Land use planning**

- ❖ Land use planning or zoning aims to control the land use so that the least number of people are living in a vulnerable area as is physically possible. By preventing new developments from occurring in hazardous areas, the size of the population decreases and hence vulnerability to the hazard event also reduces.
- ❖ In the case of hurricanes, land use planning typically involves a restriction of building on a coastline that might be prone to storm surges and also involves the building of hurricane-safe shelters underneath buildings.
- ❖ However, land use planning can be very limited as it is difficult to pinpoint exactly where to place building restrictions and hurricanes and other types of hazard can be very unpredictable.

What strategies have been designed to limit damage from hazard events and disasters?

- ❖ There are **three** ways in which a community can make adjustments to limit the damage caused by hazard events or disasters:

  1. **Modify the event**
  2. **Modify vulnerability**
  3. **Modify the losses**

Limiting damage from earthquakes

Modify the event	**Prevention:** Some scientists have attempted to modify the cause of earthquakes by pouring water into fault lines to try and lubricate the plate movement. However, these attempts have proved to be unsuccessful and it is unlikely that the technology will ever exist to modify an uncontrollable force of nature.
Modify the vulnerability	**Hazard resistant design:** Planners of areas prone to earthquakes enforce an earthquake-safe building code. Earthquake-safe buildings are usually built on hard soil to reduce vulnerability to liquefaction and often have a low centre of gravity to ensure that they are less likely to collapse. Due to the cost of constructing new buildings, retrofitting of existing buildings is preferable. Methods of retrofitting include: adding a large concrete counter-weight to the top of the building, adding rubber shock absorbers which allow the building to rock back and forth and adding cross-bracings, which helps to make a building more ductile. However, due to the expense of hazard resistant design, it is only available as an option in MEDCs. For example, 70% of buildings in Japan are either earthquake-safe or retrofitted.
	Prediction: Seismic gap theory, remote sensing of small crust movements and the direct monitoring of a pre-existing fault line can all be used to predict earthquakes (See **Page 21** for more details). Although the ability to predict an earthquake has improved, we can still only predict an earthquake within a very limited time scale.
	Preparedness: On a national scale, public participation in earthquake training schemes and practice earthquake drills in schools and the workplace (e.g. Japan) can help to provide practical information about evacuation and raise general awareness of the earthquake risk. On an individual level, making important household adjustments, such as nailing furniture to the wall, can increase preparedness for earthquakes. Fire hazards can be prevented by the use of 'smart meters', which turn off the gas supply when an earthquake of a particular magnitude occurs.
Modify the losses	**Aid:** Since earthquakes cause large immediate impacts, they tend to attract lots of emergency donations. Aid for earthquakes, in the form of rescue services and distribution of healthcare, significantly reduces the losses associated with earthquakes. However, due to the inaccessibility of some areas affected by earthquakes, aid can be difficult to deliver (e.g. the Kashmir area is highly mountainous and was difficult to reach following the 2005 earthquake).
	Insurance: Earthquake insurance is not widely available because few companies have the capital to cover the high economic costs that earthquakes cause. Take up on insurance is low due to the fact that the costs of taking out insurance can rise to a extortionate price depending on the location, type of construction and soil type.

Limiting damage from hurricanes

odify the event	**Prevention**: Prevention of hurricanes is virtually impossible and thus it is vital that people adapt to coping with them better so that they decrease their vulnerability to the hazard. However, scientists have recently experimented with seeding the outside of eye wall clouds with silver iodide, which has aimed to reduce the landfall of the hurricane by the time it reaches land. This method has had limited success, but research into hurricane prevention is on-going.
Modify the vulnerability	**Hazard resistant design:** People living in coastal areas build their roofs with stronger materials and use less glass where possible. Before a hurricane is about to occur, people might install external shutters to reduce the "explosion effect". **Prediction:** Hurricanes can be predicted by using forecasting models based on data collected from aircraft, buoys and satellite radars. The path of hurricanes can be tracked from space by satellite imaging. Hurricane prediction is becoming increasingly difficult with increasing climate variations due to global climate change (See **Page 22** for more details). **Preparedness:** Regular hurricane drills can improve hurricane preparedness. For example, Florida has introduced a new hurricane awareness programme, "Project Safeside", which implements regular hurricane drills in schools and offices. Communities can reduce their vulnerability to the storm surge of a hurricane by lining coastal flood barriers around the areas at risk. For example, sandbags were piled in front of homes on Long Beach in preparation for Hurricane Sandy in 2012. However, one limitation of this method is that putting up flood barriers can be very time consuming when the speed of onset is rapid. Afforestation is a very good preparation method as trees can reduce wind speed. For example, a green belt has been created in Bangladesh to reduce the impacts of its annual tropical cyclones. However, this method is also not a suitable short-term method of preparation.
Modify the losses	**Aid:** Although areas affected by hurricanes receive a significant amount of emergency aid, it only covers a small proportion of the losses, especially in developing regions. For example, aid only covered 10% of the losses caused by Hurricane Mitch in 1998. In the case of Hurricane Nargis, aid was initially refused and was poorly distributed when it was eventually accepted. **Insurance:** Hurricane insurance is not widely available because few companies have the capital to cover the high economic costs they cause. Take up on insurance is particularly low in LEDCs, indirectly resulting in a larger death toll.

Limiting damage from drought

dify the event	**Prevention:** Drought cannot be prevented. However, scientists have been experimenting with cloud seeding to try and modify the intensity of drought events. By injecting silver iodide into clouds, the air within the clouds starts to cool and condenses so that rainfall can occur. For example, a cloud seeding programme in Texas, USA has resulted in a 12% increase in annual rainfall, consequently reducing the intensity of meteorological drought.
Modify the vulnerability	**Prediction:** Droughts can be predicted by using drought occurrence models. These models use the data from the frequency of historic droughts and also semi-regular climatic variations caused by processes, such as the El Niño effect and sunspot activity. These models are able to give scientists information of a drought's return period. Droughts can also be predicted by using satellite imaging. From satellite imaging, scientists are able to spot clear signs of vegetation loss. **Preparedness:** The main way in which a community can prepare for drought is through conserving water supplies. Water conservation policies are targeted mainly at domestic use due to the fact that much water is needed for agriculture and consumption. Strategies include: the use of water butts to collect water for bathing, switching on the dishwasher when full and the use of water-meters with penalties for large consumption. Other strategies focus on planting drought-tolerant crops such as sorghum. This is important because it means that the impact of the drought conditions on agriculture will be reduced, hence reducing vulnerability to famine. For example, during the 2012 drought farmers in Texas planted drought-resistant crops, leading to an increase of yield of over 15%. In addition, when the first signs of drought are experienced, water from dams and reservoirs can be released to areas experiencing severe water deficit. For example, during a drought in East China in 2011, 5 billion cubic meters of water stored in the Three Gorges Dam was released to help farmers irrigate their crops.
Modify the losses	**Aid:** The most important type of relief to those in drought-affected areas is food aid. Food aid is very helpful during a period of drought because it can help minimise the number of deaths caused by malnourishment and starvation. However, large-scale distribution of food aid focuses on giving everyone an equal share instead of prioritising aid to the most vulnerable sectors of the population e.g. elderly, young children, the sick etc. **Insurance:** Drought insurance is very limited as the long duration and large extent of drought makes it an expensive option.

What are the responses to disasters on local, national and international scales?

❖ **Local scale responses:** During the event, responses at the community level might include evacuation. However, after the event responses might include, burial of dead and the distribution of basic food supplies and healthcare. Due to damages in infrastructure and lack of resources, local responses are usually limited, but are incredibly important for rapid onset hazards, such as earthquakes, as most lives are saved within the first 24 hours of the response.

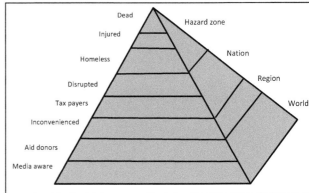

❖ **National scale responses:** During the event, responses at the national level might include the provisioning of emergency services to the areas affected by the disaster, the building of temporary shelters, the use of social media to gain international attention and support and the division of hazard risk zones (e.g.

Figure 26: Disaster impact pyramid: As the disaster begins to affect other communities, the extent of response to the disaster increases.

Montserrat exclusion zones). After the event has occurred, national scale responses might include, the distribution of compensation (human-induced only), and/or the distribution of governmental aid. The effectiveness of national scale responses is dependent on the preparedness and wealth of the affected country.

❖ **International scale responses:** During the event, international communities become aware of the impacts of a disaster via media coverage of the event. During the event, international communities become more aware of the extent of the damage caused and send emergency aid. For example, the Red Cross launched an emergency appeal for the 2010 Haiti earthquake less than 24 hours after the occurrence of the disaster. After the event, the Red Cross gave emergency shelter to over 170,000 families. Other international responses after the event might include search and rescue services and the distribution of food aid from NGOs, such as Oxfam. International scale responses are the most important in the recovery of the affected area as they help to make the previously impossible processes of rescue, rehabilitation and reconstruction possible.

What is the difference between rescue, rehabilitation and reconstruction responses?

❖ During a major hazard event or disaster, the quality of life of a community is decreased due to the large socio-economic damages. Responding to a hazard is an attempt to restore a community's quality of life to its original level.

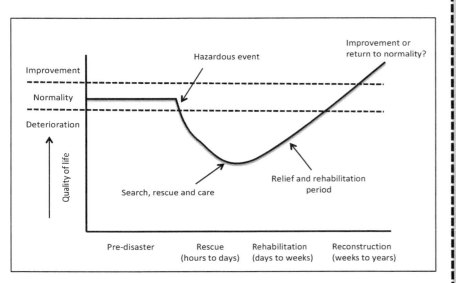

Figure 27: The Park Disaster Response model.

Case study: Responses to the Sichuan earthquake, 2008

Sichuan province is located in southeast China on the collision boundary between the Eurasian and Indian plates. On 12th May 2008, a magnitude 7.9 earthquake hit Sichuan. The earthquake lasted for two minutes and killed 70,000 people. The main impacts of this disaster included: the injury of 370,000 people, the homelessness of 5 million people and the secondary hazard of landslides adding to the already high death toll. In some rural areas, it was estimated that 80% of buildings collapsed and the total damage caused by the quake was estimated at $75 million. The impacts of this disaster were minimised due to an efficient and effective response to the event. China received a particularly large amount of international support due to the fact that this disaster occurred just a few months before the 2008 Beijing Olympic Games. Although the responses proved to be effective, public anger raised towards Chinese builders and developers for their adjustments before the event as over 9,000 school children and teachers were killed as a result of poorly built schools.

1. **Rescue:** Within hours and days of the earthquake, army forces and volunteers were searching for survivors amongst the debris. Since accessibility was difficult in the affected rural areas, 20 helicopters containing military troops and aid workers were sent to free trapped survivors and distribute emergency supplies. The majority of this immediate relief was given in the form of food supplies, clean drinking water and tents to shelter survivors from the fast-approaching monsoon rain. On 14th May, the Chinese government requested international assistance to help with the rescue efforts. For example, the Australian Red Cross raised enough funds to distribute more than 100,000 tents, safe drinking water for more than 19,000 people and sanitation facilities for 13,000 people.
2. **Rehabilitation:** In the months following the disaster, rehabilitation programmes, such as the Sichuan Earthquake Umbrella Project, set up large rehabilitation camps. Internally displaced people could go to these camps for guidance on when it would be safe to go back to their homes and they also attempted to restore agriculture in the affected rural areas through the donation of livestock and a focus on community-based activities.
3. **Reconstruction:** Reconstruction responses in the months and years following the earthquake have been highly praised. In September 2010, it was announced that all reconstruction following the 2008 disaster had been completed. Reconstruction was efficient because the Chinese government put together a reconstruction scheme shortly after the earthquake to help re-build the area by partnering villages and towns affected by the disaster with affluent provinces in other parts of the nation. The more wealthy provinces, such as Beijing and Guangdong, provided these communities with both financial and technological assistance. Large amounts of money were also spent on improving earthquake preparedness through education in primary and secondary schools. The earthquake provided an opportunity to reconstruct all public-service facilities in the affected areas with high seismic standards and modern equipment.

How are responses affected by individual and community perceptions?

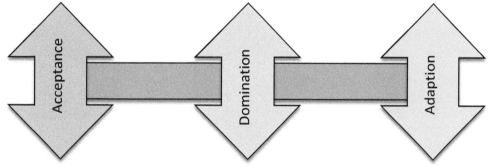

Figure 28: Response to a disaster depends on an individual's or a community's position on the hazard perception spectrum.

❖ The way in which an individual or a community responds to a hazard event or a disaster is heavily influenced by their perception of the risk posed by the hazard event (See **Page 20** for more details on hazard perception).

❖ If a hazard event is perceived as having a high risk, people are more likely to respond to it than if it is perceived as having a low risk. The risk perception of an individual is likely to be expanded to the community level.

❖ Religious individuals or communities that see hazard events and disasters as unpreventable **"acts of God"**, see them as a part of every day life and accept the associated dangers. They also believe that hazards should only be responded to for safety reasons and quite often wait for outside help. For example, some people suffering from drought in some remote and poor areas of Nigeria respond to the disaster by praying for support and rainfall.

❖ Individuals and communities who believe that hazard events and disasters can be successfully forecasted, predicted and modified by technology (e.g. hazard resistant buildings) believe that technology dominates hazards and often delay the process of response due to their 'blind faith' in a **'technological fix'**. For example, the Japanese had great confidence in their tsunami defences in protecting them from the 2011 tsunami and consequently issued less of a severe warning. However, the tsunami destroyed these defences leaving individuals to escape on their own.

❖ A community of **risk-taking individuals** would respond immediately by searching for bodies, clearing up debris and helping people evacuate from hazardous areas despite the risk of secondary hazards that could occur, such as after-shocks and flooding. For example, despite the fact that aftershocks were causing more buildings to collapse after the Haiti earthquake in 2010, many members of the local community went searching for survivors in the debris before international help was received.

❖ **Past experience** of both the individual and community also affects the responses to a disaster. For example, the eruption of the Soufriere Hills volcano in 1995 was the first eruption in Montserrat since the sixteenth century. As a result of this, perception of the hazard was low, responses to the disaster consisted of fear and panic and the response was mainly focussed on evacuation instead of the distribution of aid.

❖ Individuals and communities who perceive hazards as extreme events resulting from the interaction between human and physical processes, look at modifying the geophysical event (e.g. diverting lava flows and cloud seeding) and making the necessary adjustments (e.g. building earthquake-safe buildings and increasing hazard awareness) to reduce their vulnerability to hazards.

What factors influence the choice of adjustments and responses to a hazard event?

❖ There are many factors that influence a community's choice of adjustments and responses to a hazard event:

The **adjustments** made by a community are influenced by:

✓ **The hazard type:** It might be easier to make adjustments for a drought event than an earthquake as the speed of onset is gradual, giving a community longer to make the adjustments. Earthquakes require more long-term adjustments.

✓ **The probability of the hazard:** If the probability of an event is high then a community is more likely to invest in hazard-safe infrastructure and disaster insurance. If the probability is low, then communities may make less expensive and temporary adjustments (e.g. lining of sandbags along the coast before Hurricane Sandy).

✓ **Level of wealth:** Poor communities may not be able to afford hazard-safe infrastructure (e.g. 15% of houses in Burma were built from mud-brick) or insurance. Alternatively, adjustments in poorer communities might focus on land use planning and improving individual preparedness (e.g. attaching shutters to windows).

The **responses** to a hazard event are influenced by:

✓ **Magnitude/extent of the event:** If the hazard event has a large spatial extent and a high magnitude then more people will be affected by the event. As a result, response will occur on a much larger scale (e.g. the Texas explosion occurred on a small scale and received a limited response).

✓ **Predictability of the event:** If scientists accurately predict the event, the preparedness of the community will be high and hence emergency services will be distributed to the affected areas more quickly (e.g. Hurricane Nargis was not predicted well by scientists, so emergency services were not prepared for the severity of the event).

✓ **Level of wealth:** The distribution of governmental a is determined by how much money can be spread across the community (e.g. Haiti suffered from extreme poverty and thus had to rely on international emergency relief).

✓ **Type of government:** Some governments refuse international aid due to a fear of an uprising (e.g. responses to Hurricane Nargis were limited as the Burmese military government refused outside help)

✓ **Accessibility:** If the affected location is inaccessible then the response to the hazard event will be delayed and the main method of rescue will be airlifting people to safety (e.g. Kashmir earthquake)

Figure 29: Burma could not afford to make the necessary adjustments for Hurricane Nargis.

How important is re-assessing risk and vulnerability after a disaster?

❖ The best way to re-assess risk and vulnerability after a disaster is to observe the major issues or mistakes in previous disasters. From these observations, suitable plans can be made.

❖ When re-assessing risk and vulnerability, it is important to see how human actions and/or lack of human action intensified the impact of the hazard event. For example, following the Indian Ocean tsunami of 2004, it was found that risk was particularly high due to the removal of coral reefs and thus the tsunami had a greater impact. With this knowledge, human activities that intensify the risk of certain hazards can be restricted.

❖ It is important to learn from previous mistakes and make the necessary adjustments to ensure that the next event has less of a catastrophic impact. For example, following a drought event it should be ensured that water is transported in secure pipes so that the severity of future droughts is reduced.

❖ However, re-assessing risk and vulnerability is less important for LEDCs as they are unable to spend enough money on improving infrastructure and warning systems. Instead, all communities should focus on modifying their vulnerability to the hazard by either introducing preparation methods or by moving to a safer location. For example, following the two earthquakes in Christchurch, New Zealand in 2010 and 2011, the country's population dropped by over 10,000 people as people were moving abroad to find work in a more hazard-safe environment.

❖ Upon re-assessing risk, if alarms and warning systems proved to be ineffective (e.g. the warning time for the Tohoku tsunami in 2011 was only a few minutes) the government should aim to improve self-awareness of risk. Methods of increasing self-awareness of risk include: improving the education on hazards and investing in more effective warning systems (in countries where this is affordable). For example, upon re-assessing the risks of the 2004 Indian Ocean tsunami, a state of the art system was subsequently introduced which in turn issued a prompt warning of a potential tsunami following the 2006 Java earthquake.

B exam focus: Adjustments and responses to hazards and disasters

Study this pie chart, which shows the amount of public aid given to Haiti by nation or organization, following the 2010 earthquake.

(a.) Determine the total amount of aid given by Brazil and Saudi Arabia. **[1 Mark]**

(b.) Outline how aid can reduce vulnerability to hazard events. **[3 Marks]**

(c.) Using named examples, explain the factors that affect the choice of responses to hazard events. **[6 Marks]**

(d.) Examine the usefulness of risk assessment before deciding the strategies of adjustment and response to a hazard. **[10 Marks]**

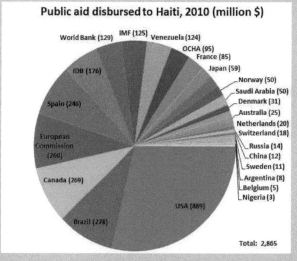

(Source: OECD database)

---

ow should I approach this question?

(a.) This question requires you to add together the figures for the amount of aid given by Brazil ($278 million) and by Saudi Arabia ($50 million). You should be able to determine that the total amount of aid given by the two countries was $328 million.

(b.) For this question, **three** brief suggestions are required. For example, suggestions could include the installation of early warning/monitoring systems, the provision of materials to limit the impacts (e.g. sandbags and metal shutters), food aid and medicines following a disaster.

(c.) You should aim to identify **three** factors (e.g. accessibility, predictability of event and wealth). You should then explain how these factors affect the choice of responses to a hazard event. Named located examples should be given to develop the explanation of each factor.

(d.) A good answer will examine cases where risk assessment has been useful (e.g. Hurricane Katrina, 2005) and examples where risk assessment has not been useful (e.g. Tohoku earthquake, 2011). Risk assessment can be examined in relation to a wide range of potential hazard events (e.g. drought, earthquakes, volcanic hazards, hurricanes and technological hazards).

End of unit checklist for optional theme D: Hazards and disasters

Sub-topic	Syllabus Statements	☹	
1. Characteristics of hazards	• Explain the characteristics and spatial distribution of the following hazards: ➢ **Either** earthquakes **or** volcanoes ➢ Hurricanes (tropical cyclones, typhoons) ➢ Droughts ➢ Any one recent human-induced (technological) hazard (explosion or escape of hazardous material) • Distinguish between the chosen hazards in terms of their spatial extent, predictability, frequency, magnitude, duration, speed of onset and effects.		
2. Vulnerability	• Explain the reasons why people live in hazardous areas. • Discuss vulnerability as a function of demographic and socio-economic factors, and of a community's preparedness and ability to deal with a hazard event when it occurs. • Explain the reasons for some sectors of a population being more vulnerable than others.		
3. Risk and risk assessment	• Examine the relationships between the degree of risk posed by a hazard and the probability of a hazard event occurring, the predicted losses and a community's preparedness for it. • Explain the reasons why individuals and communities often underestimate the probability of hazard events occurring. • Discuss the factors that determine an individual's perception of the risk posed by hazards. • Examine the methods used to make estimates (predictions) of the probability (in time and space) of hazard events occurring, and of their potential impact on lives and property. • Discuss these methods by examining case studies relating to two different hazard types.		
4. Disasters	• Distinguish between a hazard event and a disaster. Explain why this distinction is not always completely objective. • Describe the methods used to quantify the spatial extent and intensity of disasters. • Explain the causes and impacts of any one disaster resulting from a natural hazard. • Explain the causes and impacts of any one recent human-induced hazard event or disaster. • Examine the ways in which the intensity and impacts of disasters vary in space and have changed over time.		
5. Adjustments and responses to hazards and disasters	• Discuss the usefulness of assessing risk before deciding the strategies of adjustment and response to a hazard. • Describe attempts that have been made to reduce vulnerability by spreading the risk (aid, insurance) and by land-use planning (zoning). • Describe strategies designed to limit the damage from potential hazard events and disasters. • Describe the range of responses, at the community, national and international levels, during and after a hazard event or disaster. • Distinguish between rescue, rehabilitation and reconstruction responses. • Explain how these responses are affected by individual and community perceptions. • Examine the factors that affected the choice of adjustments before, and responses to, actual hazard events or disasters. • Discuss the importance of re-assessing risk, and re-examining vulnerability, following any major hazard event or disaster.		

CPI Antony Rowe
Chippenham, UK
2016-06-13 13:18